Advanced Science and Technology for Biological Decontamination of Sites Affected by Chemical and Radiological Nuclear Agents

NATO Science Series

A Series presenting the results of scientific meetings supported under the NATO Science Programme.

The Series is published by IOS Press, Amsterdam, and Springer in conjunction with the NATO Public Diplomacy Division

Sub-Series

I. Life and Behavioural Sciences	IOS Press
II. Mathematics, Physics and Chemistry	Springer
III. Computer and Systems Science	IOS Press
IV. Earth and Environmental Sciences	Springer

The NATO Science Series continues the series of books published formerly as the NATO ASI Series.

The NATO Science Programme offers support for collaboration in civil science between scientists of countries of the Euro-Atlantic Partnership Council. The types of scientific meeting generally supported are "Advanced Study Institutes" and "Advanced Research Workshops", and the NATO Science Series collects together the results of these meetings. The meetings are co-organized by scientists from NATO countries and scientists from NATO's Partner countries – countries of the CIS and Central and Eastern Europe.

Advanced Study Institutes are high-level tutorial courses offering in-depth study of latest advances in a field.
Advanced Research Workshops are expert meetings aimed at critical assessment of a field, and identification of directions for future action.

As a consequence of the restructuring of the NATO Science Programme in 1999, the NATO Science Series was re-organized to the four sub-series noted above. Please consult the following web sites for information on previous volumes published in the Series.

http://www.nato.int/science
http://www.springeronline.com
http://www.iospress.nl

Series IV: Earth and Environmental Sciences – Vol. 75

Advanced Science and Technology for Biological Decontamination of Sites Affected by Chemical and Radiological Nuclear Agents

edited by

Nelson Marmiroli

University of Parma,
Italy

Borys Samotokin

Zhytomyr State Technological University,
Zhytomyr, Ukraine

and

Marta Marmiroli

University of Parma,
Italy

Springer

Published in cooperation with NATO Public Diplomacy Division

Proceedings of the NATO Advanced Study Institute on
Advanced Science and Technology for Biological Decontamination of Sites Affected by Chemical and Radiological Nuclear Agents
Zhitomir, Ukraine
17–28 August 2005

A C.I.P. Catalogue record for this book is available from the Library of Congress.

ISBN 1-4020-5519-6 (PB)
ISBN 978-1-4020-5519-5 (PB)
ISBN 1-4020-5518-8 (HB)
ISBN 978-1-4020-5518-8 (HB)
ISBN 1-4020-5520-X (e-book)
ISBN 978-1-4020-5520-1 (e-book)

Published by Springer,
P.O. Box 17, 3300 AA Dordrecht, The Netherlands.

www.springer.com

Printed on acid-free paper

CONTENTS

PREFACE

Pollution of soils and waters by human activities is an important and widespread problem. This pollution by, organic and inorganic substances can affect individual organisms, human populations, and ecosystems, each in its own unique way. In particular former military installations, often used for weapons production and nuclear power plants represent a ongoing and substantial threat to environment and human health because of the specific pollutants that can be released: Solvents, explosives, fuels, radionuclides, heavy metals, and metalloids all have been identified in the environment around these installations. Remediation technologies for these contaminated sites have been developed based on conventional systems utilising physical and chemical treatments, such as excavation and incineration, pump-and-treat methods, ultraviolet oxidation, soil washing, etc.

These approaches are usually very expensive, and can involve the removal of large amounts of soil or water, which often leads to resources depletion and limitation in the future utilisation of the site. Recently, new decontamination approaches, based on living organisms, in particular micro-organisms and plants, have been researched and developed as a more sustainable alternative because they have lower costs, reduced environmental impacts, and increased public acceptance. Development of biological decontamination techniques has been carried out in several laboratories and research institutions worldwide, and practical applications on contaminated sites have been recorded in several countries. In the USA and Canada several commercial companies have been utilizing these approaches for years, alongside conventional techniques, with great success:; a good examples can be found is in the U.S. EPA Superfund Innovative Technology Evaluation Program. In the European Union commercialization of phytotechnologies and bioremediation technologies is still scant, even though groups of high-level scientists are actively involved in research on this very topic. Constraints to the application of biological decontamination technologies can be attributed to the legislation and to the lack of endpoint requirements for cleanup, to lack of economic considerations about costs and benefits, and to the limited knowledge of long-term environmental effects.

There is a strong need to train of new professionals in commercial or governmental initiatives, and teach them mastery the scientific background of biological decontamination and specific features of "in-field" applications. As the most relevant pollution problems are localised in the less developed countries, training per se may not suffice.

A complete capacity building policy is needed, that encompasses training, infrastructure, as well as human and financial resources necessary to implement these new technologies. Capacity building can be obtained by bringing together both scientists and experts in the field, finalized to the application of biological decontamination, in order to favor the merging of specific expertise.

These specialists are an important technical resource for governmental institutions with a mission in environmental protection, and can serve as a training resource postgraduates and post-docs who are willing to improve their professional knowledge with practical experience and education. Commercial companies and enterprises already acting or willing to move into this sector can benefit from the initiative as well.

The environmental movement is gaining acceptance and importance in many countries, particularly in those countries which have originated after the collapse of the former USSR.Many of these former Soviet Bloc countries have been left with a plethora of environmental problems arising from antiquated manufacturing processes. Built without pollution control equipment to problems arising from the careless disposal of chemicals and petroleum products, to disposal of excess military ordinance.

While there are no precise data on extent of environmental contamination in Eastern Europe, it has been estimated that the former USSR countries may contain 200,000 of these disposal sites. Western Europe, North America and other developed have made a concerted effort to identify and define their waste disposal sites, and are actively remediating them.

The NATO ASI "Advanced Science and Technology for Biological Decontamination of Sites Affected by Chemical and Radiological Nuclear Agents" held in Zhytomyr (Ukraine) on 17-28 August 2005 addressed the topics of biological decontamination of pollution related to chemical, radiological and nuclear agents. The main objectives of the initiative were: (i) to train participants for principles of scientific and technology of biological decontamination, bioremediation and phytoremediation, with particular emphasis on sites contaminated by radionuclides and chemical substances connected with explosives, ammunitions and fuels; (ii) to describe and discuss the present state-of-the-art, the latest developments, and the further advances required for commercial applications; (iii) to stimulate future interactions and collaborations in this technologically important field of study.

The purpose of the ASI was to bring together lecturers of worldwide renown in this subject and let them meet with interested stakeholders and end-users, coming from academy, research, public administration, military institutions, and private companies. Participants came from 18 countries, representing Asia, Africa, most of Europe and North America.

The ASI was organized to provide detailed, advanced, and thought provoking information about possible decontamination approaches, alternative technologies, cost effectiveness, feasibility, with the support of literature data, personal experience of lecturers, and case studies.

The chapters in this book represent the result of the lectures and of the following discussions with participants and stakeholders.

The main achievements of the school and of this book have been:

- The increased understanding of the global subject of phytoremediation-bioremediation
- The sharing of practical inside and very specialised information on subjects like decontamination of radionuclides and ammunitions or propellants
- The discussion about the need of a holistic approach with integration of different expertise: scientific, technological, legal and juridical, economical
- The understanding of the role played by all stake-holders in the cleanup process, including: scientists, economists, managers, regulators, public
- The discussion of the use of conventional vs. non conventional technologies
- The identification of faults and drawbacks in the remediation process, as well as of strengths and advantages of these processes
- The need for a policy of communication of the results achieved to stakeholders and to national and supranational regulators
- The relevance of capacity building and training,

Site characterisation procedures and related measures are addressed by Michael Pupeza (Golder Associates Srl, Italy), Oleksandr Orlov (Ukrainian Scientific-Research Institute of Forestry and Agro-Forest Amelioration, Ukraine) and Anja Hebner (BioPlanta GmbH, Germany). In particular they consider and discuss problems connected with sampling and assessment of sites contaminated by radionuclides and explosives. Pollution problems generated by the Chernobyl accident are addressed considering the contamination incurred to the forest ecosystems and the hazards to human health.

The main processes of bioremediation are addressed starting from the site conditions, analyzing the microbial features, both genetic and physiological-biochemical, in the chapter provided by Ludo Diels (Flemish Institute for Technological Research, Belgium). These basic biological considerations led further on to analyze and elucidate case studies that compared together bioremediation with conventional remediation techniques. Authors brought also their experience in technologies based on chemical reactions and construction of physical barriers, discussed alongside with examples of bioremediation. Specific cases of explosives and radionuclides decontamination were also addressed, citing European and non-European experience. New contaminants of interest are also addressed, such as the MTBE (methyl-tert-butyl-ether).

The basic biochemical mechanisms of phytoremediation are explained by Stanislaw Gawronski (Warsaw Agricultural University, Poland), considering also the differences between plant and microbial metabolism of contaminants, together with a record of the natural and cultivated plants more frequently used for decontamination. The role of genetics and genetic engineering in increasing the knowledge on detoxification processes and to produce and obtain more specific types of decontaminating plants is explained in the chapter by Nelson Marmiroli.

Practical implementation of remediation technologies is addressed starting from feasibility studies at the laboratory level, to pilot scale experiments and large scale tests. Several examples of applications are provided by researchers and private companies' representatives, both for decontamination of explosives and of radionuclides, in particular with a significant contribution by Dave Russell (Global Environmental Operations, Inc., USA), Petr Soudek and Tomas Vanek (Academy of Sciences of Czech Republic) and Christian Kunze (WISUTEC WISMUT Umwelttechnik GmbH, Germany). Constructed wetlands are prominent among successful applications and the chapters bring several examples. A straightforward analysis of case studies leads to identification of advantages and limitations of constructed wetlands technology.

The book is supplemented by the contributions of Wolf-Uwe Marr (Bundesministerium der Verteidigung, Germany) and Andrei Kozeltsev (Ministry of Natural Resources, Russian Federation) addressing the relationships between legislators and regulators on one side, and on the other side scientists and private companies acting in remediation.

The "state-of-the-art" as it emerges from the book, is that bioremediation and phytoremediation are both applied, in most cases, without a precise and detailed knowledge of all molecular mechanisms occurring within cells. This can have consequences on several aspects, and especially concerning decisions to be taken case-by-case according to site characteristics, type of pollutant(s) and endpoints to be reached.

Some main considerations deserve mentioning:

1. There is a great necessity of understanding the basic aspects of microorganism and plant physiology, biochemistry and genetics, because only from sound scientific knowledge may in future derive the possibility to drive these technologies on more applied aspects.
2. Interactions between plants and microorganisms in remediation must be studied further and with greater attention, because in many contaminated sites they both can be beneficial if applied simultaneously,, and synergy may enhance the individual site cleanup and reduce costs.. This holistic approach considers interactions in the environment, not only between organisms, but also between contaminants, and between biotic and abiotic factors.
3. Interaction of scientists and technicians with state and governmental agencies, regulators, economists, and evaluators is of paramount importance. The need to communicate and understand each other by sharing goals and objectives and develop priorities for successful implementation of biological decontamination practices.
4. Eastern European participants and scientists are convinced that phytoremediation and bioremediation can be a more sustainable solution to their environmental problems and are willing to learn more and to apply them extensively in the field.
5. The need for greater cooperation between public and private sectors can be accomplished by integrating basic academic and private technological research into a set of common social goals and priorities.
6. The promotion of a better understanding and cooperation between countries, in particular Western and Eastern European countries, favors and encourages a free access to guidance materials and basic

information on contaminated sites and previous decontamination attempts, and promotes further application of these effective and valuable techniques.

7. The support of increased capacity building of personnel, resources and infrastructures is required with a particular emphasis on young scientists and female- scientists.

ACKNOWLEDGEMENTS

The Editors, on behalf of the authors, acknowledge the financial support of NATO Science Affairs Division (EST.981075).

The Editorial Committee acknowledges the cooperation of staff members of the Institutions Zhytomyr State Technological University and University of Parma, Department of Environmental Sciences.

The editors acknowledge the professional assistance of smartenglish.com for language editing.

Nelson Marmiroli
Borys Samotokin
Steven McCutcheon
Dave Russell
Stanislaw Gawronski
Marta Marmiroli

CONTRIBUTORS

Benesová, Dagmar
Department of Plant Tissue Cultures, Institute of Organic Chemistry and Biochemistry, Czech Academy of Science, Flemingovo nám. 2, 166 10 – Prague 6, Czech Republic

Cernik, Miroslav
AQUATEST a.s., Geologická 4, Praha, Czech Republic

Diels, Ludo
Environment and Process Technology, Flemish Institute for Technological Research (VITO), Boeretang 200, BE-2400 Mol, Belgium

Ducco, Francesco
Golder Associates Europe Ltd., Via Antonio Banfo 43, Torino, Italy

Gawronska, Helena
Laboratory for Basic Sciences in Horticulture, Faculty of Horticulture and Landscape Architecture, Warsaw Agricultural University, Nowoursynowska 159, 02-776 Warsaw, Poland

Gawronski, Stanislaw W.
Laboratory for Basic Sciences in Horticulture, Faculty of Horticulture and Landscape Architecture, Warsaw Agricultural University, Nowoursynowska 159, 02-776 Warsaw, Poland

Gerth, André
BioPlanta GmbH, Deutscher Platz 5, D-14103 Leipzig, Germany

Gigli, Paola
Golder Associates Europe Ltd., Via Antonio Banfo 43, Torino, Italy

Greco, Marco
Golder Associates Europe Ltd., Via Antonio Banfo 43, Torino, Italy

Hebner, Anja
BioPlanta GmbH, Deutscher Platz 5, D-14103 Leipzig, Germany

Kiessig, Gunter
WISUTEC Wismut Umwelttechnik GmbH, Jagdschänkenstr., D-09117 Chemnitz, Germany

Kuechler, Annette
WISUTEC Wismut Umwelttechnik GmbH, Jagdschänkenstr., D-09117 Chemnitz, Germany

Kovalchuk, Andriy
Zhytomyr State Technological University, 103, Cherniakhovsky str., Zhytomyr 10005, Ukraine

Kozeltsev, Andrey. L.
Minister of Natural Resources of Russian Federation, 4/6, B. Gruzinskaya St., Moscow, 123995, Russian Federation

Kunze, Christian
WISUTEC Wismut Umwelttechnik GmbH, Jagdschänkenstr., D-09117 Chemnitz, Germany

Levitsky, Vyacheslav
Zhytomyr State Technological University, 103, Cherniakhovsky str., Zhytomyr 10005, Ukraine

Lookman, Richard
Environment and Process Technology, Flemish Institute for Technological Research (VITO), Boeretang 200, BE-2400 Mol, Belgium

Marmiroli, Marta
Department of Environmental Sciences, University of Parma, Viale G.P. Usberti 11A, 43100 Parma, Italy

Marmiroli, Nelson
Department of Environmental Sciences, University of Parma, Viale G.P. Usberti 11A, 43100 Parma, Italy

Marr, Wolf-Uwe
Bundesministerium der Verteidigung, WV IV 2, Postfach 1328, D53003 Bonn, Germany

McCutcheon, Steven
Faculty of Engineering, Driftmier Engineering Center, University of Georgia, Athens, GA 30602, USA

Micalizio, Emiliano
ERG Petroli, Via Brancati 60, 00144 Rome, Italy

Orlov, Oleksandr
Polis'ky Branch of Ukrainian Scientific-Research Institute of Forestry and Agro-Forest Amelioration, 38 Prospect Myru, Zhytomyr, 10004, Ukraine

Podlipna, Radka
Department of Plant Tissue Cultures, Institute of Organic Chemistry and Biochemistry, Czech Academy of Science, Flemingovo nám. 2, 166 10 – Prague 6, Czech Republic

Pupeza, Michael
Golder Associates Europe Ltd., Via Antonio Banfo 43, Torino, Italy

Russell, David L.
Global Environmental Operations, Inc., 4642 Warrior Trail, SW., Lilburn, GA, 30047, USA

Samolyuk, Igor
Zhytomyr State Technological University, 103, Cherniakhovsky str., Zhytomyr 10005, Ukraine

Samotokin, Borys
Zhytomyr State Technological University, 103, Cherniakhovsky str., Zhytomyr 10005, Ukraine

Soudek, Petr
Department of Plant Tissue Cultures, Institute of Organic Chemistry and Biochemistry, Czech Academy of Science, Flemingovo nám. 2, 166 10 – Prague 6, Czech Republic

Vakriková, Zuzana
Department of Plant Tissue Cultures, Institute of Organic Chemistry and Biochemistry, Czech Academy of Science, Flemingovo nám. 2, 166 10 – Prague 6, Czech Republic

Valenová, Šárka
Department of Plant Tissue Cultures, Institute of Organic Chemistry and Biochemistry, Czech Academy of Science, Flemingovo nám. 2, 166 10 – Prague 6, Czech Republic

Vaněk, Tomáš
Department of Plant Tissue Cultures, Institute of Organic Chemistry and Biochemistry, Czech Academy of Science, Flemingovo nám. 2, 166 10 – Prague 6, Czech Republic

Yanchuk, Valentyn
Zhytomyr State Technological University, 103, Cherniakhovsky str., Zhytomyr 10005, Ukraine

PART 1

SITE CHARACTERISATION

METHODOLOGY AND APPROACH TO ENVIRONMENTAL SITE INVESTIGATIONS OF LARGE INDUSTRIAL SITES IN ORDER TO OBTAIN OPTIMAL COST/BENEFIT RATIOS: CASE STUDY OF THE ISAB – ERG PETROLI REFINERY AT PRIOLO GARGALLO (SR), ITALY

M. PUPEZA[1*], P. GIGLI[1], F. DUCCO[1], E. MICALIZIO[2]
1 Golder Associates s.r.l., Torino, Italy*
2 ERG Petroli, Rome, Italy

*To whom correspondence should be addressed.

Abstract- The process of environmental site characterisation implies a set of investigations to establish a solid description of the site. This then represents the basis for the decision-making process underpinning the design of a feasible and sustainable environmental rehabilitation program. Given the financial constraints related to the containment or remediation of a large site such as the one discussed herein, it is imperative that the site characterization is driven by a cost/benefit analysis. The characterization of a large industrial site first requires the appropriate choice, from a pool of potential technologies, of the optimal combination of investigation methods and data evaluation procedures. Based on this approach, it is possible to satisfy both the criteria of maximum efficiency and most favorable cost/benefit ratio. Most of the methods used at the target site are common in the area of oil-field exploration. The specific site conditions rendered these highly effective, and meant that intrusive investigation could be minimized. Only ~40 monitoring wells were installed over a period of about four months, at depths varying from 25 to 90m (equivalent to ~1600 drilled meters), instead of the ~350 borings (~12,000 drilled meters) which would have been mandated by current Italian legislation. The approach used allowed for a sufficiently detailed characterization of the site; and specifically led to the precise definition of two active sources of impact that were promptly eliminated, and the focusing of direct, intrusive, investigation only to areas of highest interest, thus reducing both intervention time and overall cost. A feasibility study and a site remedial

N. Marmiroli et al. (eds.), Advanced Science and Technology for Biological Decontamination of Sites Affected by Chemical and Radiological Nuclear Agents, 03–14.

design are currently underway. The installation and operation of dedicated geophysical data loggers in monitoring wells allows for the continuous monitoring of contamination over time, and may also drive the development of novel equipment able to identify accidental sub-surface spills and trigger alarms whenever these occur.

Keywords: Environmental site characterisation; tiered approach; detailed site characterization; direct and indirect investigations; maximum efficiency and best cost/benefit ratio

1. Introduction

An environmental site characterization comprises the set of investigations which form the basis for a solid knowledge of the site. This is turn constitutes the necessary foundation for the decision-making process leading to a feasible and sustainable environmental rehabilitation of the site. The results of such work must include a characterization of the site sub-surface, and an assessment of the presence and dynamics of any contamination in order to define the best approach for impact containment and site remediation.

As required by Italian law, the environmental site characterization was undertaken of the ISAB Refinery, owned by Erg Petroli of Priolo Gargallo (SR), Italy and included in the Priolo "national interest site" by decree of the Italian Environment Ministry. A cost/benefit analysis was chosen based on a combination of traditional methods (geological survey, geo-diagnostic investigations) and advanced technologies (down-hole geophysical survey) rather than on simpler criteria (investigation point grid). This latter invasive statistical approach, more easily managed by the control bodies, is often prescribed by legislation, even though it cannot provide an exhaustive environmental assessment at reasonable cost. Hence the choice was made to apply a tiered approach throughout the environmental characterization work, combining indirect and invasive methods. This approach was discussed with the Italian Authorities and subsequently approved by the Italian Environment Ministry as being suitable for characterizing the ISAB site.

Geophysical techniques are especially useful for investigating large industrial sites potentially contaminated by hydrocarbons in order to focus and minimize invasive (direct) investigations (drilling). Direct investigations are nevertheless essential, as these allow investigators to calibrate the indirect investigation and extrapolate from data obtained from specific points to a larger area, thus providing a greater coverage of the area under investigation. The integrated assessment of the results overall allows investigators to accurately plan further investigations, both direct (boreholes and monitoring wells) and indirect (non-invasive, targeted geophysical tests).

The objective of this article is to show how to apply a tiered approach, making broad use of non-invasive methods, where possible, especially for large industrial sites with complex sub-surface characteristics, maximizing efficiency in terms of time- and cost-effectiveness, and determining all factors necessary for the rapid design of containment measures and for long-term site remediation.

2. Methodology

The environmental site characterization approved by the Italian Environment Ministry involved indirect investigation methods serving to focus and minimize the direct investigations. The site investigations comprised:

- a detailed geological survey of the superficial rock formation in the area of the refinery
- a superficial geophysical survey using geo-electrical measurements to create resistivity profiles of the sub-surface (covering approximately 25,000 lineal meters)
- stratigraphic drilling and installation of monitoring wells
- geophysical down-hole logs (resistivity, natural γ-ray, spontaneous potential) and installation of sensors and electrical connections for periodic monitoring
- interpretation and validation of geophysical data on the basis of point data (stratigraphy and down-hole logs)
- increasing the density of investigation points with monitoring wells in the areas of concern

The indirect investigations were carried out in the early phase of work, in parallel with the first direct investigations, and required approximately two months to complete. The results of the early direct investigations allowed the calibration of the results of the indirect investigation. The overall data obtained at this stage from the field operations permitted a preliminary environmental site characterization, guiding the successive planning of more extensive direct investigations. The point data from the direct investigations could thus be extended, and the need for further investigations could be determined.

The results of indirect and direct investigations were constantly validated against one another and reinterpreted where necessary on the basis of reciprocal results, and field / laboratory data, regarding both information in the sub-surface (e.g., state and degree of fracturing of bedrock, lithological variations, groundwater table level, salinity, seawater intrusion) and its contamination (e.g., presence of LNAPL).

2.1. GEOLOGICAL SURVEY

The refinery has been in operation since the early 1970s, and occupies a total area of approximately 350ha. It is located near the sea on a rocky substrate comprising fractured and tectonicized, locally karsic limestone and sandstone, overlying a volcanic formation. The sub-surface hydrogeology comprises a carbonate aquifer that flows seawards through a sandstone formation subject to significant saltwater intrusion. The rock complex is highly heterogeneous, producing quite varied environmental characteristics within individual geological and hydrogeological units.

The geological survey produced a preliminary characterization of the following sub-surface elements:

- lithology and geological complexes
- bedrock fracturing (spacing, persistence and filling)
- influence of fracture systems on surface water and groundwater flow and the local hydrogeological system
- seawater intrusion areas and natural contamination containment phenomena
- high vulnerability areas
- preferential contaminant transport pathways (percolation and advection) advection)

Regarding the development of a structural model of the refinery sub-surface, it was observed that the discontinuities present in the area can be grouped into three distinct systems. With the exception of stratification (primary discontinuity), all three of these systems are represented by joints (fractures where the two opposing faces do not shift with respect to one another), shear joints (where the two faces do shift with respect to one another) and faults (fault planes or zones, marked by cataclasis, rubble and mylonite with clear evidence of shifting of the two faces).

In order to determine the significance of the surface discontinuities and assess their persistence and connections at depth, useful information was obtained from a combined analysis of structural data (joints and faults of different scales, comparing the discontinuities recognized on the ground) and photographic data, and from a cross analysis with the results of geophysical measurements and geo-diagnostic investigations carried out in the area of the refinery. It should be emphasized that the integrated interpretation of geology and geophysics allowed investigators to gain an understanding of the complex hydrological system in the area.

2.2. GEOPHYSICAL (INDIRECT) INVESTIGATIONS

The geophysical investigations comprised:

- Tomographic resistivity profiling (by means of sub-surface resistivity profiles, and
- Down-hole geophysical logs (resistivity measurements by means of down-hole electrodes, reflectance at frequencies between 50 and 150MHz and natural γ-ray logging)

Electrical resistivity tomography was employed to obtain site sub-surface resistivity profiles to depths ranging from 30 to 40 meters below ground level.

Electrical resistivity profiling using multi-array tomography is an innovative methodology growing out of classic geo-electrical measurements, and was used to achieve high spatial resolution. The system can reconstruct the spatial distribution of sub-surface resistivity in two or three dimensions, the resolution depending on the number of electrodes and the distance between them. The electrical investigation was carried out by means of multi-electrode profiling, collecting resistivity data with a mixed Wenner and dipole-dipole device. The data were then integrated for processing via tomographic inversion to produce profiles of real resistivity. The length of the profiles investigated along the main axes in certain cases exceeded 1km. The guarantee measurement continuity "roll along" procedure was used, which involves running out the string to half its length, and repeating along the overlapping section the Wenner and dipole-dipole measurements to ensure the continuity of the profile.

The down-hole logs involved:

- Electric resistivity profiles were determined by means of down-hole electrode arrays. The instrumentation used for superficial tomography was also used for down-hole electrode measurements. Over 1,000 down-hole electrodes were installed to validate the surface data and give greater precision with respect to the assessments made from the surface, making them also usable for future monitoring. Via the reconstruction of tomographic profiles (see above) and the down-hole electric logs, the resistivity measurements allowed investigators to reconstruct the real resistivity of the sub-surface. This parameter is a function of the natural characteristics of the sub-surface and of any fluid contained in its solid matrix. In the case in question, resistivity measurements allowed the investigation team to detect structural anisotropies (lateral contacts, fractured zones with any fine filling, etc.) and hydrocarbon accumulation zones.

- Acoustic reflectance achieved by the measurement of the intensity of the reflection of a wave emitted at frequencies between 50 and 150MHz. A greater proportion of reflectance indicates widespread heterogeneity, as for example in fill, in coarse soil types, in highly fractured rock, or sometimes in cases of organic liquids present in the pore space.
- The measurement of natural radioactivity with a γ-ray detector determined the activity coefficient (the quantity of γ-rays emitted), which is a specific characteristic of each type of sediment or rock. Specifically, a high γ-ray count indicates the unmistakable presence of a clay matrix and thus constitutes useful information for the synthesized, integrated interpretation of this and other measurements performed.
- Natural γ-ray spectroscopy logs were introduced in the early 1970s as a common and inexpensive measurement for oil exploration, although they had been studied since the 1950s. A log of the total natural radioactivity can usually be made both in the open hole and through the casing. During drilling operations the log is also used for correlation between wells, for depth correlation between open and cased holes, and for depth correlation between other logging runs. For this last use the γ-ray log was the first nuclear well log and was introduced in the late 1930s.

In order to use this technology for environmental purposes, the data obtained through the geophysical investigation had to be calibrated on the basis of the structural-geological survey and on core data obtained from the boreholes so that the environmental characteristics of the investigated sub-surface could be attributed to them.

2.3. DIRECT (INVASIVE) INVESTIGATIONS

On the basis of the indirect investigations and in compliance with the prescriptions of the Italian Environment Ministry, environmental investigations were carried out at the refinery using continuous core recovery drilling (water was used as the drilling fluid for a total of approximately 2,700 linear meters of drilling. A total of 50 Lugeon permeability tests were carried out during drilling operations. The boreholes were capped with 4” piezometer tubes to serve as monitoring wells. Drilling operations took approximately four months with a daily progress rate of approximately 15m per drill rig per day.

The following activities were carried out:

- 14 shallow boreholes to a maximum depth of 8m
- 68 deep boreholes were fitted out as monitoring wells to depths ranging from 25 to 88m
- 25 supplemental deep boreholes were equipped as monitoring wells to determine the exact nature and extension of the contamination detected previously in the various refinery zones
- soil samples were collected at 1m intervals for on-site Head Space Analysis (HSA)
- A collection of 450 soil samples was taken for laboratory analysis, selected on the basis of the HSA, with particular focus on fracture points filled with fine-grained materials (silt-clay)

The boreholes were sited on the basis of the accumulating results of the indirect investigations. Specifically, the following criteria were applied:

- The interpretation of homogeneous and background data from the indirect investigations
- The assessment of anomalies (localized areas of limited extension) recorded during the indirect investigations and not interpretable on the basis of the structural-geological survey

It was thus possible to extend point data to homogeneous areas and to determine with a higher degree of precision the nature of the anomalies revealed, whether due to structural anisotropy or to the presence of organic fluids in the subsurface.

2.4. INTEGRATED DATA ANALYSIS

An integrated interpretation was carried out using a simultaneous, synthetic analysis of the data obtained from each individual investigation method, both for indirect (geophysical and structural-geological surveys) and for direct investigations (borehole drilling and monitoring well tests). Particularly in the case of the geophysical investigations, this approach allowed investigators to make an environmental interpretation of the physical data obtained (e.g., resistivity) both for zones with uniform results and for those exhibiting anomalies. The anisotropies influence the dynamics of fluids (groundwater and LNAPL) in the sub-surface.

By way of illustration, four examples of integrated data analysis are presented relative to the following sub-surface characteristics:

- CASE 1 – Determination of lithological contacts (vertical and horizontal) and depth to groundwater table (integrated evaluation of geophysical data, structural-geological survey data and data obtained through direct investigations)
- CASE 2 – Evaluation of the importance of structural anisotropies (fractures, faults, cataclastic bands) and their persistence (horizontal and vertical) measured at the surface (structural-geological survey) and their mapping (integrated evaluation of geophysical data and structural-geological data)
- CASE 3 – Evaluation of the presence and the areal extension of organic fluids (LNAPL) floating on the aquifer (integrated evaluation of geophysical data and data collected from direct investigations involving monitoring wells)
- CASE 4 – Determination of areas characterized by salt water intrusion (integrated evaluation of geophysical data and electrical conductivity data from direct investigations in monitoring wells)

2.5. AREAL COVERAGE METHOD

Generally a statistical approach involves a reconstruction of the environmental sub-surface framework through the interpolation of point data obtained via direct methods (note that Italian legislation stipulates one monitoring well per 10,000m^2 surface area). For the particular site, given its high sub-surface complexity, we have hypothesized that the point data obtained from the drilling of one monitoring well is significant only within a radius of 3 - 5m. The use of indirect geophysical methods allowed an extension from the information obtained from one or more drilling points, located within a swath 3 - 5m wide along a geophysical investigation profile, to achieve significantly greater areal coverage. The optimization of the direct investigation points necessary for the proper reconstruction of the sub-surface characteristics is nevertheless a function of the complexity of the sub-surface, both in terms of natural characteristics (geological and hydro-geological) and the state of contamination (number and extension of contaminant plumes). A choice was therefore made to locate an average of one investigation point per every 230m of the geophysical investigation profile. In the zones of greater complexity, whether natural or related to the state of contamination, the investigation points were sited every 23m.

2.6. EFFECTIVENESS OF THE METHOD

The general analysis of the investigation results demonstrates the effectiveness of the method, most notably regarding the highly heterogeneous natural characteristics of the sub-surface of the site:

- the areal coverage of the information obtained through the investigation methodology employed at the ISAB site was estimated at 200,000m^2
- the overall investigated area of 3,500,000 m^2 required 107 direct investigation points, compared to the 350 that would have been required if a statistical approach had been used
- a total of 2,700 linear meters of perforations were carried out, as opposed to the theoretically estimated 12,000m, achieving a reduction of over 75%
- the investigation methodology resulted in the coverage of an additional area of about 170,000m^2 (200,000m^2, as compared to 27,000m^2)
- the methodology used allowed the investigation team to highlight two areas in the refinery showing anomalies related to the presence of LNAPL in the sub-surface, to focus the environmental investigation and to implement emergency containment measures (elimination of primary sources of contamination)
- in terms of the time needed to produce similar results, the overall time requirement for the investigations (about four months) is significantly less than is needed for the 350 investigation points prescribed by Italian law (estimated at 18 months). This aspect is of great value in cases where the swift implementation of emergency containment measures, designed on the basis of investigation results, is necessary
- the integrated analysis of the structural-geological survey and the geophysical investigation provided a detailed delineation of the hydro-geological model of the sub-surface, by identifying the structural anisotropies in the bedrock as well as those that govern the dynamics of sub-surface fluids (whether groundwater or LNAPL)
- the detailed reconstruction of the hydro-geological make-up of the sub-surface allowed a focus on direct investigations on the most significant and sensitive points in terms of groundwater contaminant transport
- regarding the total number of anomalies recorded via the geophysical investigation, approximately 90% are linked to sub-surface structural anisotropies (fractures, faults, vertical or lateral lithological contacts), 5% are related to salt water incursions, and 5% to sub-surface contamination by LNAPL

Detailed knowledge of the geological and hydro-geological make-up of the sub-surface, obtained exclusively via the methodology described here, is essential for highly heterogeneous rocky substrates. The structural make-up governs the dynamics of sub-surface fluids, identifying high vulnerability zones and zones that receive water, and therefore contamination, within the system. The identification of such zones allows the identification of the most sensitive areas and the design of suitable containment measures, excluding non-essential areas and thus saving both time and money. These areas must also be taken into account when carrying out a risk assessment for the site as part of the decision-making process to determine which measures to adopt to safeguard the environment.

3. Cost analysis

A cost analysis and its comparison with traditional investigation methodologies was an essential element in evaluating the performance of the methodology applied at the ISAB site. The complete site investigation cost was €775,000. A traditional methodology (350 investigation points and approximately 12,000 linear meters of drilling) of the same area would have cost approximately €2.8M. The siting of direct investigation points using conventional methodology (i.e., not guided by preliminary indirect investigations) would have required an increase in drilling point density in the more significantly contaminated areas. For the methodology we applied, the density variations were determined during the preliminary investigation phase, focussing particularly on the two areas where LNAPL contamination was revealed. It is evident that, given the structural complexity of the sub-surface of the site, a purely conventional approach, even if supported by a structural-geological survey, would not have been able to produce a detailed structural and hydro-geological model of the sub-surface with the same precision as achieved by the methodology described here.

In assessing costs, the time factor should not be overlooked. The method followed required approximately six months of effective field work (two months for indirect investigations and four for direct investigations), compared to an estimated 18 months needed for the conventional methodology, generating attendant savings of internal resources for managing operations and subcontractors. The time factor was also essential in determining the containment measures which, if implemented swiftly, will serve to reduce the overall cost of site remediation.

4. Conclusions

For the environmental characterization of the ISAB Refinery owned by Erg Petroli of Priolo Gargallo (SR), Italy, the choice was made to use an advanced approach and complex technologies, as opposed to simpler conventional methods, which would not have been able to produce an exhaustive environmental characterization at a sustainable cost. The application of technologies typically used for hydrocarbon exploration adapted to the environmental context has allowed for the characterization of the site using a lesser number (about 40) of geo-diagnostic boreholes and monitoring wells installed to depths ranging from 25 - 90m below the ground surface, for a total of about 1,600 linear meters of perforation (requiring about four months to complete), as opposed to the 350 boreholes as would have been required if a statistical approach had been taken (about 1 borehole per ha) for an estimated total of 12,000 linear meters of drilling. The density of investigation points was increased in the contaminated areas to delineate the extent of the sources of LNAPL. Approximately 30 supplementary boreholes and monitoring wells (1,100m of perforation) were installed in these areas.

The approach applied has permitted greater coverage of the site area, the identification of two active sources of contamination and their timely elimination. This focussing of the direct investigations has achieved significant savings in both time (investigation work and data interpretation) and cost. The tiered approach allows investigators to proceed with a closer spacing of direct investigation points and to intervene quickly, where necessary, with containment measures. The cost savings were evaluated in comparison to a methodology which would not have produced similarly valuable results either in terms of environmental site characterization or the identification of contaminated areas. The assessment and design of emergency containment measures was concentrated only on those areas where these measures were truly necessary, improving the effectiveness of the work (time and operational method), while at the same time reducing costs.

The installation of fixed systems for down-hole geophysical measurements will permit a more refined monitoring of the environmental state of the sub-surface and the development of automatic alarm systems to ensure swift response in the event of critical changes in the quality of the sub-surface. The overall reduction of investigation time has led to the installation of a containment system about one year earlier than would have been possible, had a classical approach been employed.

RADIATION CONTAMINATION OF FOREST ECOSYSTEMS INVESTIGATION: FIELD EXPERIMENTS, MODELING AND SIMULATION

OLEKSANDR ORLOV[1]
Polis'ky Branch of Ukrainian Scientific-Research Institute of Forestry and Agro-Forest Amelioration, 38 Prospect Myru, Zhytomyr, 10004, Ukraine

ANDRIY KOVALCHUK, VYACHESLAV LEVITSKY, IGOR SAMOLYUK, VALENTYN YANCHUK
Zhytomyr State Technological University, 103, Cherniakhovsky str., Zhytomyr 10005, Ukraine

Abstract- Forests are a critical source of internal exposure to radiation pollution in the form of forest food (mushrooms and berries). Forests are more heavily polluted than open agricultural areas, because of the higher transfer coefficients of radionuclide migration. To evaluate the influence of forest food on the internal exposure dose to the population living in heavily contaminated forested regions, two population groups were identified, based on their usage of forest food. For the internal exposure dose calculation over the post-Chernobyl catastrophe period, the modeling of ^{137}Cs migration in the forest ecosystem was taken into account.

Keywords: ^{137}Cs, Specific Activity, Total Activity, Forest Ecosystems, Layers of Vegetation, Modeling, Simulation

[1] To whom correspondence should be addressed. Oleksandr Orlov Polis'ky Branch of Ukrainian Scientific-Research Institute of Forestry and Agro-Forest Amelioration, 38 Prospect Myru, Zhytomyr, 10004, Ukraine, e-mail: station@zt.ukrpack.net

N. Marmiroli et al. (eds.), Advanced Science and Technology for Biological Decontamination of Sites Affected by Chemical and Radiological Nuclear Agents, 15–44.

1. Introduction

Mixed coniferous/broadleaved forest occupies about 70% of the area of the northern part of Ukraine and plays a very important landscape role in the region. This ecosystem is characterized by a great diversity of ecological conditions and variability in its spatial distribution, by a complex bio-geochemistry (cycling) of the major nutritional elements and artificial contaminants, including man-made radionuclides (Melin and Wallberg, 1991, Orlov et al. 1999), by a continuously growing canopy and a multi-layered vegetation. There is a rich species composition in the lower layers of the forest, including about 600 species of grasses and dwarf-shrubs, 300 of macromycetes, 200 of mosses and 250 of lichens. Among these are about 30 species of edible mushrooms, 10 of berries and 50 of wild medicinal plants. The consumption of these 'forest gifts' by the local population in the presence of radioactive contamination can result in a significant increase in their internal exposure to irradiation.

The level of radioactive contamination in the forest can be much higher than in open country (up to 10 times greater), but the density of radionuclide at forest ground level is very variable, because of the tree stands. ^{137}Cs ground deposition in the zone affected by the Chernobyl accident can vary up to 50 fold. This variability makes it difficult to predict the specific activity of radionuclides in particular components of the forest ecosystem, and forces a probabilistic approach.

Artificial rehabilitation of the forest contaminated by radionuclides is impossible (Pushkariov et al., 2000; Dolin and Sobotovich, 2000; Dolin, 2000). Thus these ecosystems will remain a source of radiological danger for many years, especially where forest food products (wild berries and medicinal plants, edible mushrooms, game animals) are consumed. A significant proportion of radionuclide activity returns annually to the soil with litter fall, and radionuclides are reabsorbed through the roots (Tikhomirov and Shcheglov, 1997; Shcheglov, 1999). The various layers of the forest vegetation are characterized by very significant sorption capacities of certain radionuclides via the aerial route, but some species cannot absorb via the roots (e.g., green mosses and lichens). By investigating patterns of the distribution of ^{137}Cs activity in various layers of the forest ecosystem, a basis can be set for a prognostic mathematical modeling of the migration of ^{137}Cs.

The tasks of this research were:

- The analysis of the spatial variability of density of ^{137}Cs at ground level in the forest

- The analysis of ^{137}Cs distribution, and the identification of key components of its geochemistry
- A description of the vertical distribution of ^{137}Cs in the soil
- The measuremen tof ^{137}Cs content in forest biota on experimental plots
- Definition of the role of 'forest gifts' as a source of internal radiation exposure of the local population in Ukrainian Polissya

2. Experimental plots

The research was carried out in the 'obligatory settling out zone' (Zhytomyr region, Polissya of Ukraine) in summer 2004, using three experimental plots which represented distinct pine wood forest ecosystems. The experimental plots were selected on the basis of approximate equal age of the tree canopy (50 years), and similarity of species composition and type of growth conditions. Each 1.0ha plot was defined using standard methods (Yunatov, 1964).

Plot 1: Cladonio-Pinetum type: Soils – dune sands. Tree canopy *Pinus sylvestris* of mean height 6.7m and mean diameter 9.3cm. Undergrowth consisting of *Chamaecytisus ruthenicus* (Fisch. ex Woł.) Klaskova. Grass–dwarf-shrub layer rare, with a total projective cover of 10-12%, mainly *Corynephorus canescens* (L.) P.Beauv (5-7%), *Thymus serpyllum* L. (1-3%) and *Calluna vulgaris* (L.) Hull (3-5%). Lichen layer represented by dense epigeous lichens with a projective cover of 85-90% and dry biomass of 2.0kg/m^2, mainly *Cladonia mitis* (Sandst.) Ruoss (60-65%), and epiphytic lichens (mainly *Hypogymnia physodes* (L.) Nyl). On the driest dune sites there were fragments of moss cover (*Polytrichum piliferum* Hedw. (3-5%) and *Pohlia nutans* (Hedw.) Lindb.); macromycete layer represented mainly by symbiont mushroom *Lactarius rufus* (Scop). Fr., *Xerocomus badius* Fr. and *Paxillus involutus* (Batsch.) Fr.

Plot 2: Dicrano-Pinetum type: Soils - hills of sandy dune, weakly soddy-podzolic sand; tree canopy *P. sylvestris* of mean height 18.5m and mean diameter 17.4cm; undergrowth absent. Grass–dwarf-shrub layer rare, mainly boreal species *C. vulgaris*, *Vaccinium vitis-idaea* L. and *Melampyrum pretense* L. Green moss layer dense, >1.2kg/m^2 dry weight, consisting of two main species – *Pleurozium schreberi* (Brid). Mitt. (40-50%) and *Dicranum polysetum* Sw. (40-45%). Lichen layer fragmental, including *C. mitis*, *C. rangiferina* and others. Macromycete layer consisting of 15 species including the main edible species of mushrooms in the region – *Boletus edulis, Rozites caperata, X. badius* and others.

Plot 3: Molinio-Pinetum: Soils - soddy-podzolic sandy-loam, with an average annual depth of ground water level about 1.3m. Tree canopy *P. sylvestris* of mean height 23.5m and mean diameter 22cm, and some *Betula pubescens* Ehrh. Undergrowth rarefied consisting of *Frangula alnus* Mill. Grass–dwarf-shrub layer total projective cover 60-75%, mainly *Vaccinium myrtillus* L. (50-60%), *V. vitis-idaea* (5-10%) and *C. vulgaris* (1-5%). Moss layer projective cover about 80-95% consisting of green mosses *P. schreberi* (40-50%) and *D. polysetum* (30-45%). Lichen layer epiphytic *Hypogymnia physodes* and *Pseudoevernia furfuracea.* Macromycetes layer including about 10 species, most commonly *R. caperata*, *Amanitopsis fulva* and *Russula paludosa* (L.) Kuntze, whose fruitbodies are responsible for the bulk of the macromycete biomass.

3. Methodology

For the determination of the distribution of ^{137}Cs in the experimental plots, it was necessary to measure both the weight of each component and the specific activity of radionuclide. The main characteristics of the tree canopy were described using standard methods (Anuchin, 1977). The parameters of an average model tree were calculated for three main stages of trunk thickness. Three model trees were cut on each experimental plot, each representing a certain stage of trunk thickness. Each was divided into its separate organs and tissues and weighed. Branches were classified by diameter into thin (diameter <5mm) and thick (>5mm) (Myakushko, 1978)) and samples from the crown of each tree were taken in proportion to their weight; the upper, mean and lower and further parts were combined into one integrated sample for each tree. The trunk was sawn into 1m sections, and the bark removed. Wood and bark were weighed separately. The sample of trunk wood was taken from a height of 1.3m. A combined needle sample was taken from branches of different sections of the crown, proportionally to their weight; samples of annual shoots were taken in a similar fashion. The weight of thin roots (diameter <2mm) and thick roots (>2mm) from each model tree was determined. Root samples were collected uniformly from the soil profile. Above- and below-ground biomass measurements of the grass–dwarf-shrub layer were based on five replicates of a 25m^2 area. The roots of vascular plants were washed free of soil in the laboratory. Epigeious lichens were analysed as five replicates of 1m^2, and macromycetes from an area of 100m^2. Epiphytic lichens were sampled from the model trees before their cutting. The green moss layer was sampled from experimental plots in automorphic conditions as five replicates of 500cm^2. Separate fractions of forest litter (recent, semi-decomposed and decomposed) and mineral soil samples were taken from a

500cm^2 square; the depth of each mineral soil sample was 2cm, sampled down to a depth of 30cm.

All samples were dried for 72 hours at 80°C, and dry weights were calculated. Dried samples were milled to spectrometrically measure the specific activity of ^{137}Cs. The standard error of specific activity was in the range 10-20%. Statistical analysis used the software package MS Excel.

4. Results and discussion

To describe the spatial variability of density of ^{137}Cs ground deposition in the forest ecosystem, a specific programme was focussed on an experimental plot in the Molinio-Pinetum ecosystem. The 'enclosed squares' method was applied, in which each new square overlaps the previous, larger, square, and the sampling intensity decreases, in this case reducing from every 100m to every 0.1m (Fig. 1). The lowest ground density of ^{137}Cs was 111kBq/m^2 (equivalent to 3Ci/km^2) and the highest about 1000kBq/m^2. ^{137}Cs hotspots were distributed fairly uniformly over the plot, governed by the structure of the tree canopy. Some important conclusions arising from this study were that the ground deposition of ^{137}Cs is characteristically variable, and that its amplitude is independent of sampling method. Overall, the difference in radionuclide content of the various species resulting from the spatial variability of the soil radio-contamination is around one to two orders of magnitude.

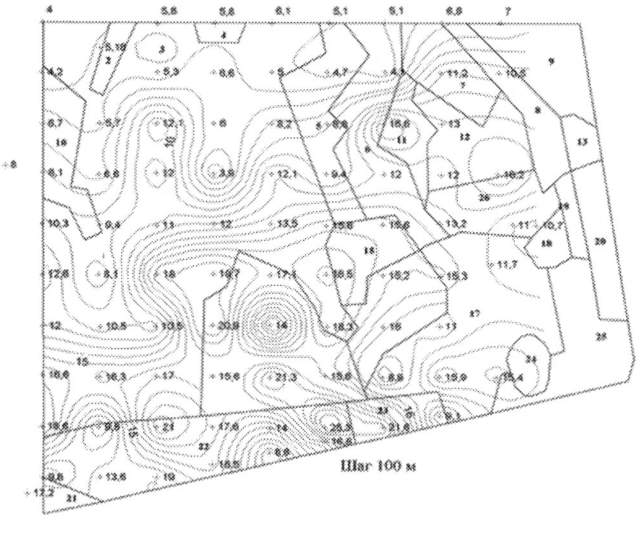

Figure 1a. Variability of density of 137Cs ground deposition on experimental plot of Molinio-Pinetum investigated with different step of sampling (sampling step 100 m).

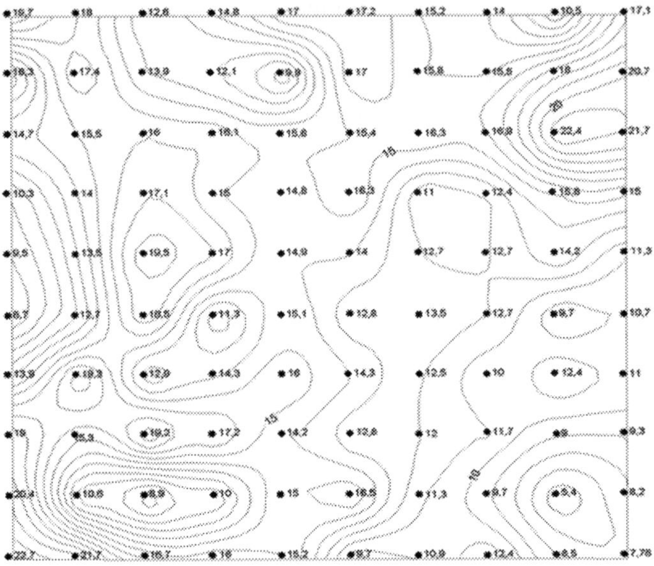

Figure 1b. Variability of density of 137Cs ground deposition on experimental plot of Molinio-Pinetum investigated with different step of sampling (sampling step 10 m).

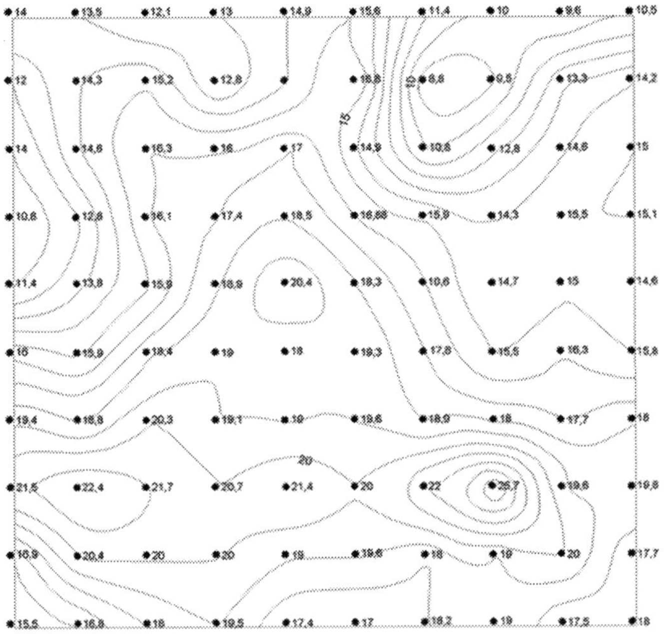

Figure 1c. Variability of density of 137Cs ground deposition on experimental plot of Molinio-Pinetum investigated with different step of sampling (sampling step 1 m).

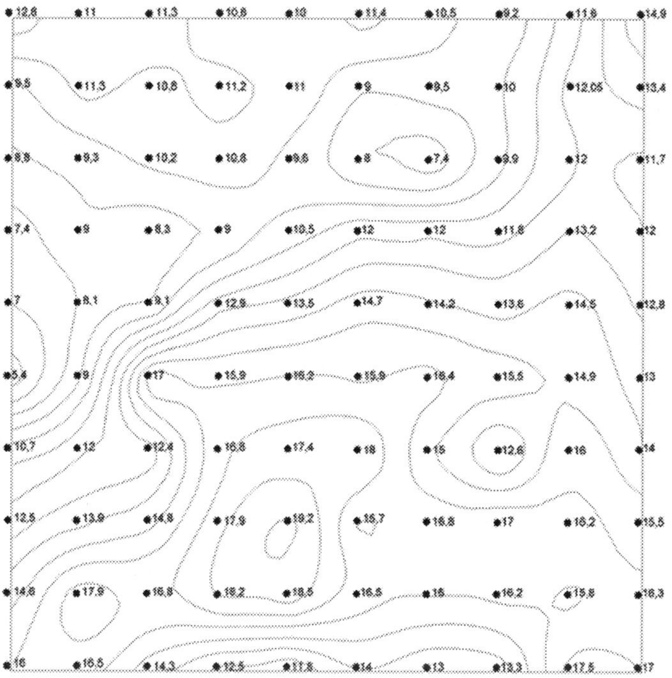

Figure 1d. Variability of density of 137Cs ground deposition on experimental plot of Molinio-Pinetum investigated with different step of sampling (sampling step 0,1 m).

Note: density of ^{137}Cs ground deposition is done in Ci/km^2.

The forest soil type is an important factor in determining the accumulation of radionuclides by plants and fungi, and therefore influences the distribution of ^{137}Cs in the forest ecosystem (Mamikhin et al., 1997). Among the most important peculiarities of the soils of Ukrainian Polissya (in common with all soils in the Coniferous-Broadleaved forests of the boreal type) is the presence of a substantial layer of forest litter, consisting mainly of pine needles and moss residuals. The decomposition time of such litter material is from six to 12 years, so this layer provides a long-term reservoir of radionuclides. Each soil layer differs from others not only by its colour but also in its major physical and chemical properties. The commonest soils of the region are sands and sandy-loams on fluvio-glacial deposits, poor in both humus (0.2-0.5%) and mineral elements, especially K^+, Ca^{2+}, Mg^{2+}. These soils are also characterized by their high acidity, with the pH of the soil solution ranging from 3 to 5.5. The rhizoshpere is typically shallow (15-20cm in depth), with most of the roots and mycelium being restricted to the superficial soil layer, where the highest level of

Table 1. The distribution of 137Cs in three contrast forest ecosystems.

Component of ecosystem	Mass, kg/ha	^{137}Cs specific activity, kBq/kg	^{137}Cs activity, MBq/ha	Part of ^{137}Cs activity,%
Forest ecosystem Cladonio-Pinetum (A_1), density of ^{137}Cs ground deposition – 492,35 kBq/m^2 (13,31 Ci/km^2)				
Tree canopy (Pinus sylvestris)	38834	19,81	769,231	11,89
Wood	23674	7,12	168,564	2,61
Bark external	1969	8,10	15,955	0,25
Bark internal	175	21,91	3,834	0,06
Annual needles	2177	91,00	198,107	3,06
Needles of the 2-nd year	3304	28,57	94,395	1,46
Annual shoots	164	111,08	18,217	0,28
Twigs thick	5861	32,39	189,838	2,94
Twigs thin	1510	53,20	80,332	1,24
LAYER OF UNDERGROWTH OF TREES (PINUS SYLVESTRIS)	11,97	6,91	0,083	0,001
LICHEN LAYER	20743,7	83,721	762,396	11,79
Sublayer of epigeious lichens	20735,78	36,75	762,024	11,774
Cladonia mitis	7550	34,58	261,079	4,04
Cladonia rangiferina	4795	37,50	179,813	2,78
Cladonia crispata var. crispate	2285	37,54	85,779	1,33
Cladonia crispata f. elegans	888	36,55	32,456	0,50
Cladonia gracilis	2765	40,20	111,153	1,72
Cladonia furcata	510	32,14	16,391	0,25
Cladonia subulata	247,5	32,80	8,118	0,13
Cladonia uncialis	812,5	38,68	31,428	0,49
Cladonia turgida	405	33,50	13,568	0,21
Cladonia cornuta	450	39,02	17,559	0,27
Cladonia digitata	1,84	170,00	0,313	0,005
Cladonia chlorophaea	7,83	166,44	1,303	0,02
Cladonia bacillaris	3,35	143,90	0,482	0,01
Cladonia rei	5,20	185,98	0,967	0,01

Table 1. Continued.

Cladonia verticillata	1,53	153,04	0,234	0,004
Cladonia pyxidata	8,03	172,00	1,381	0,02
Sublayer of epiphytic lichens	7,92	46,97	0,372	0,006
Hypogymnia physodes	4,60	44,50	0,205	0,003
Pseudevernia furfuracea	2,23	42,50	0,095	0,001
Cladonia chlorophaea	0,6	64,56	0,039	0,0006
Cladonia rei	0,35	68,98	0,024	0,0004
Cladonia bacillaris	0,14	69,04	0,097	0,0001
MOSS LAYER	1,56	94,166	0,147	0,002
Pohlia nutans	0,28	106,00	0,030	0,0005
Ceratodon purpureus	0,13	112,02	0,015	0,0002
Pleurozium schreberi	0,33	92,02	0,030	0,0005
Dicranum polysetum	0,31	90,00	0,028	0,0004
Polytrichum piliferum	0,44	88,00	0,039	0,001
Brachythecium oedipodium	0,07	81,00	0,006	0,0001
GRASS–DWARF-SHRUB LAYER	49,59	9,161	0,454	0,01
Corynephorus canescens	14,58	2,69	0,039	0,001
Carex ericetorum	2,93	3,02	0,009	0,0001
Calluna vulgaris	5,05	29,78	0,150	0,002
Festuca ovina	2,33	2,01	0,005	0,0001
Arctostaphylos uva-ursi	13,15	17,68	0,232	0,004
Solidago virgaurea	6,16	1,72	0,011	0,0002
Hieracium umbellatum	5,39	1,50	0,008	0,0001
LAYER OF MACROMYCETES	10,95	1013,524	11,098	0,172
Sarcodon imbricatus	1,65	1950,00	3,218	0,05
Hydnum rufescens	0,11	999,89	0,110	0,002
Xerocomus chrysenteron	0,12	1262,14	0,515	0,002
Coltricia perennis	0,13	72,50	0,328	0,0001
Amanita porphyria	0,82	400,00	0,328	0,01
Amanita spissa	0,24	383,00	0,092	0,001
Cortinarius semisanguineus	0,9	2198,11	1,978	0,03
Paxillus involutus	2,26	1025,80	2,318	0,04
Cantharellus cibarius	0,63	337,83	0,213	0,003
Tricholoma flavovirens	3,03	596,43	1,807	0,03
Tricholoma portentosum	0,48	586,15	0,281	0,004
Boletus edulis	0,11	620,00	0,068	0,001
Lactarius rufus	0,13	1332,67	0,173	0,003

Table 1. Continued.

Cortinarius mucosus	0,13	1214,16	0,158	0,002
Xerocomus badius	0,21	916,86	0,193	0,003
SOIL	5510200	0,894	4923,474	76,145
Forest litter	14800	58,705	868,840	13,43
Ol	1200	17,40	20,880	0,32
Of+Oh	13600	62,35	847,960	13,11
Mineral soil layers	357200	1,40	4054,634	63,035
0-2 cm	355200	6,17	2191,584	33,89
2-4 cm	357200	1,40	500,080	7,73
4-6 cm	336400	0,75	252,300	3,90
6-8 cm	375200	0,57	213,864	3,31
8-10 cm	350800	0,48	168,384	2,60
10-12 cm	368800	0,36	132,768	2,05
12-14 cm	383600	0,34	130,424	2,02
14-16 cm	357600	0,29	103,704	1,60
16-18 cm	372000	0,21	78,120	1,21
18-20 cm	365600	0,19	69,464	1,07
20-22 cm	357000	0,19	67,830	1,05
22-24 cm	375600	0,18	67,608	1,05
24-26 cm	410400	0,11	45,144	0,70
26-28 cm	382000	0,06	22,920	0,35
28-30 cm	348000	0,03	10,440	0,16
TOTAL	–	–	6466,883	100,00
Forest ecosystem Dicrano-Pinetum (A_2), density of ^{137}Cs ground deposition – 384,41 kBq/m^2 (10,39 Ci/km^2)				
Tree canopy (Pinus sylvestris)	63126	10,48	661,298	12,27
Wood	51570	6,11	315,093	5,85
Bark external	3843	7,66	29,437	0,55
Bark internal	117	42,88	5,017	0,09
Annual needles	936	68,68	64,004	1,19
Needles of the 2-nd year	693	31,17	21,601	0,40
Annual shoots	198	112,56	22,287	0,41
Twigs thick	4086	26,27	107,339	1,99
Twigs thin	1683	57,35	96,520	1,79
LAYER OF UNDERGROWTH OF	18	8,12	0,146	0,003

Table 1. Continued.

TREES (PINUS SYLVESTRIS)				
LICHEN LAYER	169,91	40,70	6,915	0,133
Sublayer of epigeious lichens	165	40,76	6,725	0,12
Cladonia mitis	52	41,28	2,147	0,04
Cladonia rangiferina	72	41,14	2,962	0,05
Cladonia gracilis	41	39,42	1,616	0,03
Sublayer of epiphytic lichens	4,91	38,68	0,190	0,003
Hypogymnia physodes	3,05	37,72	0,115	0,002
Pseudevernia furfuracea	1,86	40,26	0,075	0,001
MOSS LAYER	14325	35,15	503,467	9,35
Pleurozium schreberi	5490	34,24	187,978	3,49
Dicranum polysetum	6270	36,67	119,921	4,27
Hylocomium splendens	2565	33,36	85,568	1,59
GRASS–DWARF-SHRUB LAYER	14,2	63,63	0,904	0,02
Calluna vulgaris	4,5	49,50	0,223	0,004
Vaccinium vitis-idaea	5,2	53,00	0,276	0,01
Convallaria majalis	2,3	69,46	0,160	0,003
Melamryrum pretense	2,2	111,54	0,245	0,005
LAYER OF MACROMYCETES	25,1	1325,40	33,228	0,62
Paxilus involutus	1,2	1634,03	1,961	0,04
Amanita citrine	0,59	887,00	0,523	0,01
Amanita pantherina	0,39	992,00	0,387	0,01
Amanita muscaria	0,3	953,00	0,286	0,01
Amanita porphyria	0,8	968,18	0,775	0,01
Cantharellus cibarius	0,4	601,40	0,241	0,004
Suilus variegates	0,81	1627,00	1,318	0,02
Boletus edulis	0,41	616,60	0,253	0,005
Lactarius rufus	5,6	1440,40	8,066	0,15
Lactarius helvus	3,81	1350,00	5,144	0,10
Russula emetica	1,21	726,10	0,879	0,02
Russula decolorans	0,39	594,60	0,232	0,004
Cortinarius mucosus	0,75	890,10	0,668	0,01
Rozites caperata	3,95	1929,20	7,620	0,14
Xerocomus badius	4,46	1093,50	4,877	0,09
SOIL	5313600	0,723	3844,115	77,61

Table 1. Continued.

Forest litter	*45600*	*22,20*	*1012,412*	*18,79*
Ol	2400	10,10	24,240	0,45
Of	24400	36,97	902,068	16,74
Oh	18800	4,58	86,104	1,60
Mineral soil layers	5268000	0,538	2831,703	58,82
0-2 см	223600	2,71	605,956	11,25
2-4 см	323200	1,16	374,912	6,96
4-6 см	349600	1,48	517,408	9,60
6-8 см	360800	1,08	389,664	7,23
8-10 см	360000	1,10	396,000	7,35
10-12 см	317200	1,07	339,404	6,30
12-14 см	362400	0,54	195,96	3,63
14-16 см	367200	0,38	139,536	2,59
16-18 см	359600	0,26	93,496	1,74
18-20 см	386400	0,12	46,368	0,86
20-22 см	375600	0,05	18,780	0,35
22-24 см	379200	0,04	15,168	0,28
24-26 см	364000	0,04	14,560	0,27
26-28 см	361600	0,03	10,848	0,20
28-30 см	377600	0,03	11,328	0,21
TOTAL	–	–	5387,493	100,00
FOREST ECOSYSTEM MOLINIO-PINETUM (B_3) density of ^{137}Cs ground deposition – 597,25 kBq/m^2 (16,14 Ci/km^2)				
TREE CANOPY (PINUS SYLVESTRIS)	160351	7,29	1168,259	15,17
Wood	134086	5,28	707,974	9,19
Bark external	10103	7,93	80,117	1,04
Bark internal	336	27,25	9,156	0,12
Annual needles	2584	53,66	138,657	1,80
Needles of the 2-nd year	1767	21,13	37,337	0,48
Annual shoots	467	43,03	20,095	0,26
Twigs thick	7460	13,08	97,577	1,27
Twigs thin	3548	21,80	77,346	1,00
LAYER OF UNDERGROWTH OF TREES (PINUS SYLVESTRIS)	18,44	7,44	0,137	0,002
LICHEN LAYER	12,66	46,91	0,594	0,01
Sublayer of epigeious lichens	*1,31*	*46,92*	*0,088*	*0,010*

Table 1. Continued.

Cladina chlorophaea	0,84	66,70	0,056	0,001
Cladina rei	0,47	69,00	0,032	0,0004
Sublayer of epiphytic lichens	11,35	44,35	0,505	0,0014
Hypogymnia physodes	6,82	43,42	0,296	0,004
Pseudevernia furfuracea	4,53	46,20	0,209	0,003
MOSS LAYER	14243	35,72	508,715	6,61
Pleurozium schreberi	5800	35,41	205,378	2,67
Ptilium crista-castrensis	484	37,12	17,966	0,23
Dicranum polysetum	6990	36,09	252,269	3,28
Hylocomium splendens	503	34,81	17,509	0,23
Sphagnum nemoreum	466	33,46	15,592	0,20
GRASS–DWARF-SHRUB LAYER	546,77	28,96	15,834	0,21
Vaccinium vitis-idaea	21,2	29,32	0,622	0,01
Vaccinium uliginosum	326,06	29,12	9,495	0,12
Vaccinium mirtillus	36,56	32,98	1,206	0,02
Ledum palustre	158,63	27,80	4,410	0,06
Molinia caerulea	4,32	23,68	0,102	0,001
LAYER OF MACROMYCETES	18,94	1804,56	34,178	0,44
Rozites caperata	7,65	1992,60	15,243	0,20
Paxilus involutus	1,45	2196,57	3,185	0,04
Lactarius rufus	1,33	1697,00	2,257	0,03
Lactarius helvus	1,47	1662,83	2,444	0,03
Russula paludosa	2,12	1900,03	4,028	0,05
Cortinarius mucosus	2,26	2089,99	4,723	0,06
Amanitopsis fulva	2,66	863,57	2,297	0,03
SOIL	4982800	1,199	5972,472	77,56
Forest litter	*72800*	*59,394*	*4323,848*	*56,14*
Ol	1600	10,80	17,280	0,22
Of	36800	103,50	3808,800	49,46
Oh	34400	14,47	497,768	6,46
Mineral soil layers	4910000	0,336	1648,624	21,41
0-2 см	144000	0,139	200,160	2,60
2-4 см	252400	0,48	121,152	1,57
4-6 см	264400	0,32	84,608	1,10
6-8 см	291200	0,76	221,312	2,87
8-10 см	288000	0,69	198,720	2,58
10-12 см	300400	0,62	186,248	2,42
12-14 см	351600	0,36	126,576	1,64

Table 1. Continued.

14-16 см	364400	0,25	91,100	1,18
16-18 см	391600	0,26	101,816	1,32
18-20 см	396800	0,25	99,200	1,29
20-22 см	382800	0,23	88,044	1,14
22-24 см	377600	0,10	37,760	0,49
24-26 см	365200	0,09	32,868	0,43
26-28 см	364400	0,09	32,796	0,43
28-30 см	375200	0,07	26,264	0,34
TOTAL	–	–	770,090	100,00

radio-contamination is present. This factor largely explains the intensive accumulation of ^{137}Cs by plants and fungi in forest ecosystems (Orlov et al., 2000). These soil properties allow for the high accessibility of man-made radionuclides for plant roots and fungal mycelia. Thus, even in the situation of relatively low levels of radioactive contamination, the ^{137}Cs content of forest plants and the fruit-bodies of mushrooms can exceed by 10-100 fold those observed in agricultural ecosystems. Table 1 presents the specific activities of ^{137}Cs in the various components of the forest ecosystem. It is clear that the ^{137}Cs content of the forest litter is higher than that of the mineral soil. However, the mass of the forest litter per unit area is significantly less than that of the mineral soil, so the distribution of ^{137}Cs between these parts of forest soil profile is rather specific, and depends heavily on the plant community present (Fig. 2).

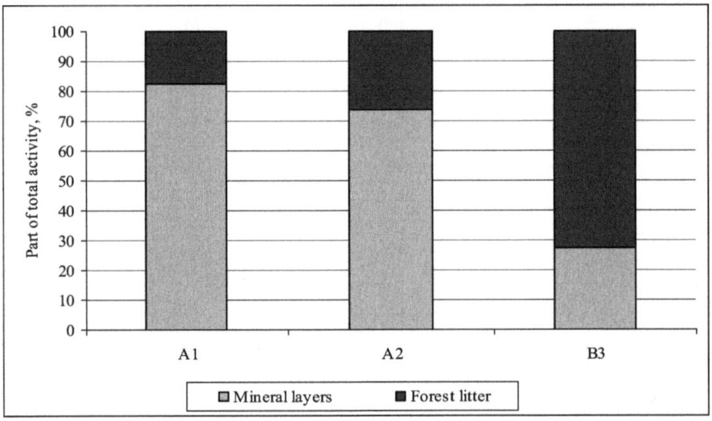

Figure 2. The distribution of ^{137}Cs between forest litter and mineral thickness of forest soil (to the depth of mineral soil 30 cm) in different types of forest habitats.

In the Cladonio-Pinetum ecosystem, the forest litter layer (depth 0.5-1cm) retains only 18% of the ^{137}Cs activity of the soil, the remainder having migrated into the mineral soil. In the Dicrano-Pinetum ecosystem, the forest litter is deeper (5-10cm) and has significant mass per unit area, allowing it to retain 26% of the ^{137}Cs activity. Finally, in the Molinio-Pinetum forest, the litter layer can reach 20-30cm, and its retention of ^{137}Cs is as high as 72% of the total soil activity. The forest litter itself consists of three layers - O_l, the new litter; O_f,, the semi-decomposed (fermented) litter; and O_h, the fully decomposed (humified) litter. The vertical distribution of ^{137}Cs content through these three layers is dependent on soil type, but commonly ^{137}Cs specific activity is relatively lower in O_l than in the other layers. A sharp increase in activity is characteristic for O_f,, falling away in O_h (Fig. 3). Thus, in the Cladonio-Pinetum, the ^{137}Cs specific activity was 17.4kBq/kg in O_l, and 62.3kBq/kg in O_f+O_h (these two layers are difficult to define separately in this ecosystem); in the Dicrano-Pinetum, O_l activity was 10.1kBq/kg, O_f 37.0kBq/kg and O_h 4.6kBq/kg; while in the Molinio-Pinetum the respective levels were 10.8kBq/kg, 103.5 kBq/kg and 14.5kBq/kg. The semi-decomposed forest litter is not only characterized by the highest levels of ^{137}Cs specific activity, but also the largest mass per unit area, resulting in the observed distribution of ^{137}Cs activity in the forest litter (Fig. 3). In the Cladonio-Pinetum, due to fragmentary presence of O_l on the lichen surface and its low ^{137}Cs content, its contribution to the retention of this radionuclide is only 2.4%, with the remainder of the activity present in O_f+O_h. In the Dicrano-Pinetum, O_l contributes 2.4%, O_f 89.1% and O_h 8.5%. The analogous values for the Molinio-Pinetum are O_l 0.4%, O_f 88.1% and O_h 11.5%.

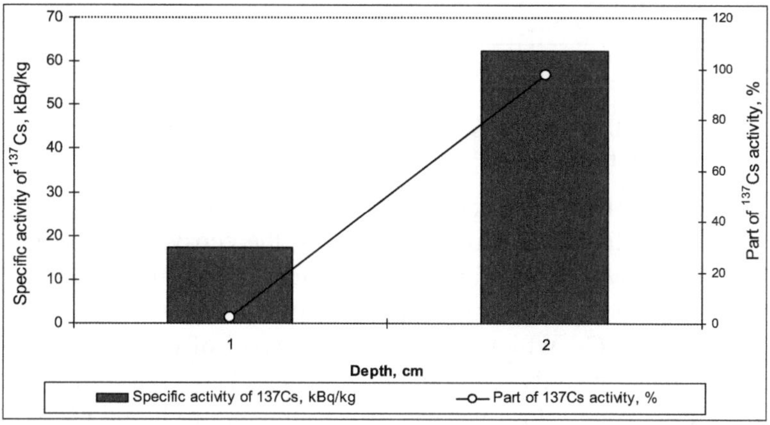

Figure 3a. The distribution of ^{137}Cs specific activity and sum activity in different fractions of forest litter in forests: A – A_1.

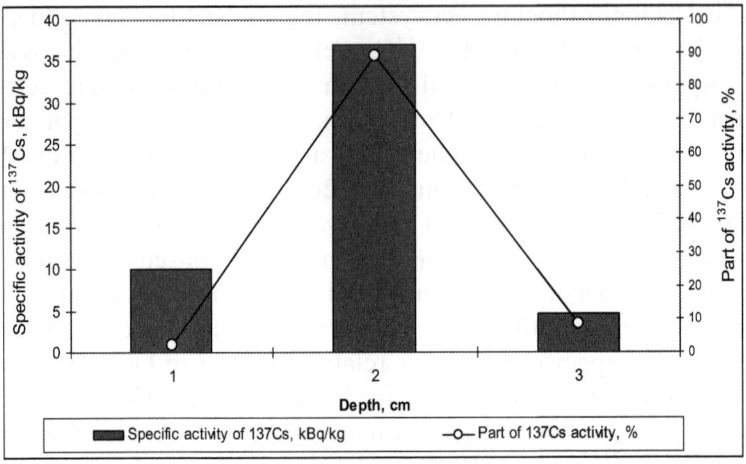

Figure 3b. The distribution of ^{137}Cs specific activity and sum activity in different fractions of forest litter in forests: B – A_2.

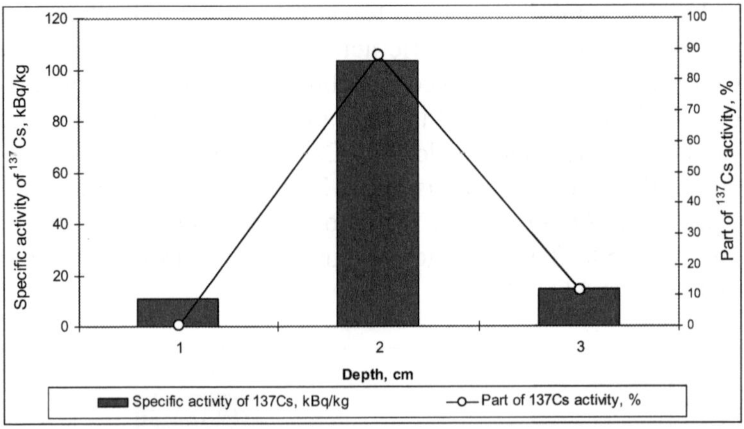

Figure 3c. The distribution of ^{137}Cs specific activity and sum activity in different fractions of forest litter in forests: C – B_3.

The process of forest litter decomposition is one of the key processes leading to the redistribution of ^{137}Cs among the ecosystem components. It is affected by two major factors: the rate of decomposition, which is dependent on the composition of the litter; and the time since the introduction of ^{137}Cs, which determines the level of decomposition of the uppermost, most heavily contaminated layer. The distribution of ^{137}Cs between the separate litter layers is important, since O_f and O_h are critical for the roots of many undergrowth plant species (such as *V. myrtillus, V. vitis-idaea*, *V. uliginosu* and *L. palustre*) and fungal saprotrophs (*Clitocybe*,

Lepista, *Mycena* species) and symbionts (particularly the edible species *X. badius*, *X. chrysenteron* and *Lactarius* spp.). The decrease in ^{137}Cs specific activity with increasing depth of the mineral soil has one particularly important ecological consequence: the coincidence of species-specific rhizospere with a variable load of radioactivity generates inter-specific differences in ^{137}Cs accumulation. For authomorphous soil types, a decrease in radionuclide content with depth is common, but the gradient of this decrease is quite variable (Fig. 4). In some soddy-weakly-podzolic soils, the vertical distribution of specific activity is quite uniform, but in others, secondary peaks of activity were associated with the presence of heavier soil at various depths (6-12cm; 16-20cm). However, overall, the vertical distribution of ^{137}Cs activity was comparable in all the soils analysed.

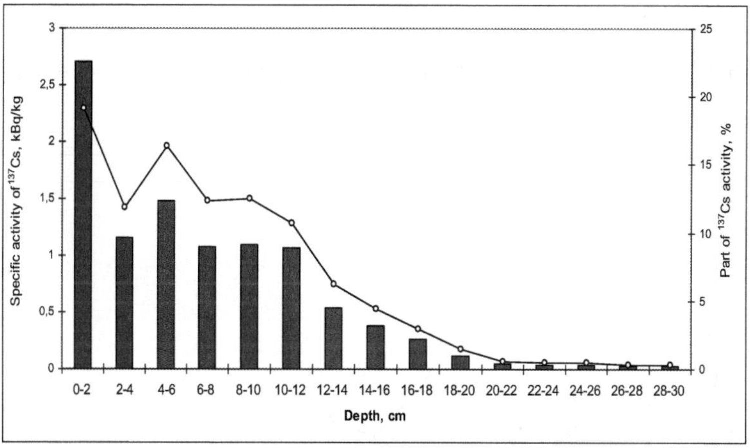

Figure 4a. Decreasing of ^{137}Cs specific and sum activity in mineral layers of different forest soils A_1.

Our experiments have allowed a description of the distribution of ^{137}Cs activity, as well as the mass of each ecosystem component per unit area (Table 1). Some important outcomes are:

- Significant interspecific differences in the ^{137}Cs accumulation within a single layer of vegetation in each ecosystem
- An exponential decrease in the ^{137}Cs specific activity of soil, as soil depth increases
- The dependence of total ^{137}Cs activity in a given layer on the amount of biomass present, and on the mode of radionuclide accumulation (via the roots or via the aerial parts of the plants)

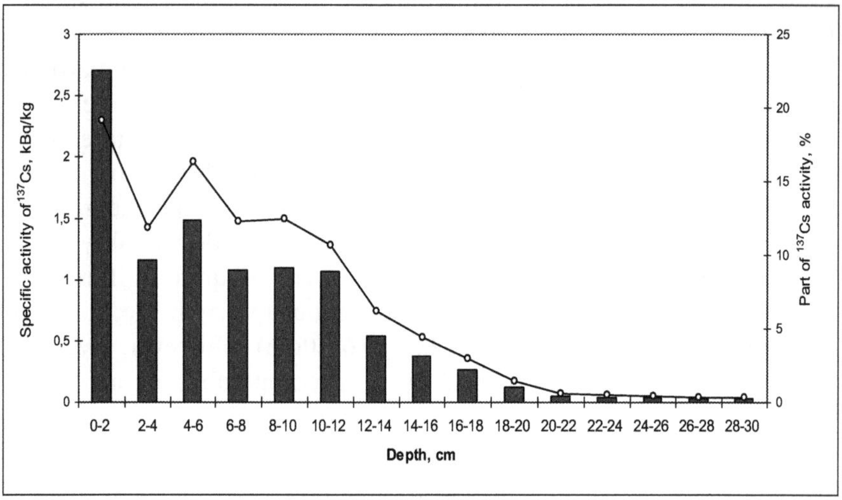

Figure 4b. Decreasing of ^{137}Cs specific and sum activity in mineral layers of different forest soils A_2.

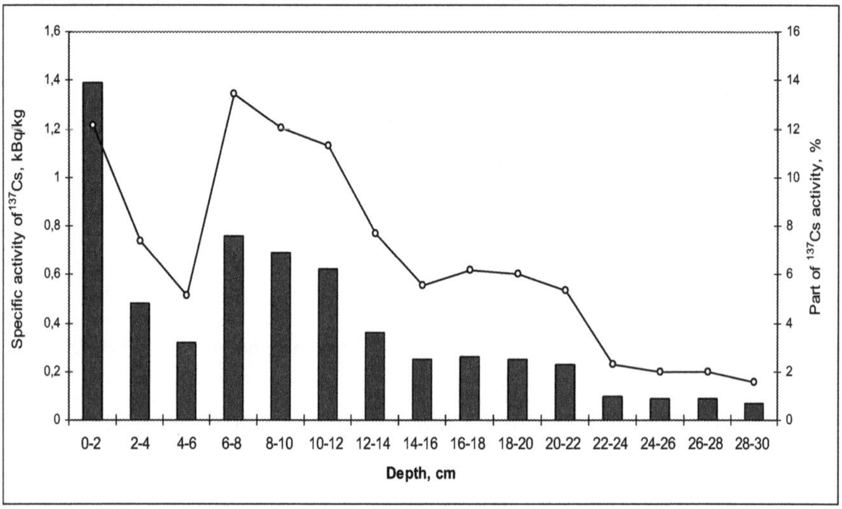

Figure 4c. Decreasing of ^{137}Cs specific and sum activity in mineral layers of different forest soils B_3.

The specific activity of ^{137}Cs is variable within a given experimental plot for all layers of the plant community present. For example, in the Cladonio-Pinetum ecosystem, the grass–dwarf-shrub layer species differed in their ^{137}Cs specific activity by almost 20 fold (Fig. 5), ranging from about 30kBq/kg (*C. vulgaris*) to 1.5kBq/kg (*H. umbellatum*). The same applies to the macromycete layer, where the variation was about 30 fold, ranging from 2198kBq/kg (*C. semisanguineus*) to 72kBq/kg (*C. perennis*)

(Fig. 6). The most important component of the forest ecosystem is the tree canopy, as this largely determines the intensity of ^{137}Cs cycling in the whole ecosystem. The range in ^{137}Cs specific activity of the components of this layer of forest vegetation is displayed in Fig. 7. Unlike the narrow variation in ^{137}Cs ground deposition measured in the experimental plots, the content of this radionuclide in the components of the pine trees were markedly variable, especially in the active growing components such as the internal bark, the annual shoots and the annual needles. Less variation was present in the wood and external bark. The importance of these results lies in the fact that the lowest level of tissue resistance to irradiation corresponds to those parts of the plant carrying the highest radionuclide content, with clear implications for radiological damage to the pine stands. Fig. 6 shows that ^{137}Cs accumulation is highest in the annual shoots, followed by the annual needles, the thin twigs, the internal bark, the second year needles, the thick twigs, the external bark, and finally the wood. This pattern is important for our future understanding of the role of each of these components in the redistribution of radionuclide in the tree canopy. The same analysis was applied for all the vegetation layers, and we have presented, as an example, the Cladonio-Pinetrum system (Figs. 8 and 9). The average weighted values of ^{137}Cs specific activity (calculated by dividing the total activity of each component by its biomass per unit area) are illustrated in Fig. 10. Thus the highest average weighted ^{137}Cs specific activities were present in the macromycete layer, followed by the moss and lichen layer, the grass–dwarf-shrub layer, and finally the tree canopy and undergrowth of trees. The radionuclide content among these components within a plot varied by three to four orders of magnitude.

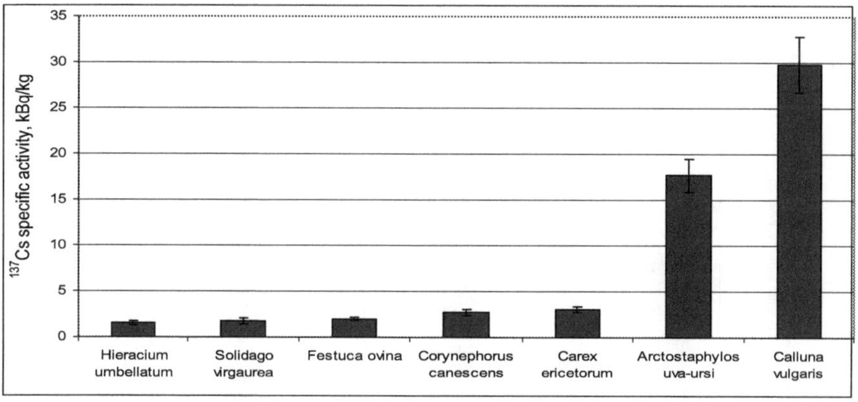

Figure 5. ^{137}Cs content in plants of grass–dwarf-shrub layer of cenosis Cladonio-Pinetum (A_1).

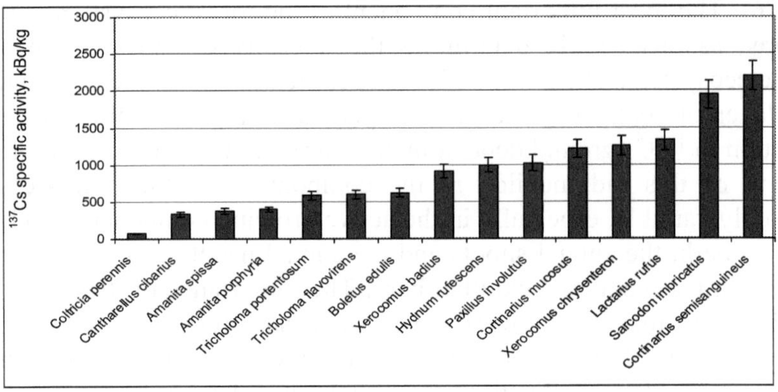

Figure 6. ^{137}Cs content in fruitbodies of macromycetes in cenosis Cladonio-Pinetum (A_1).

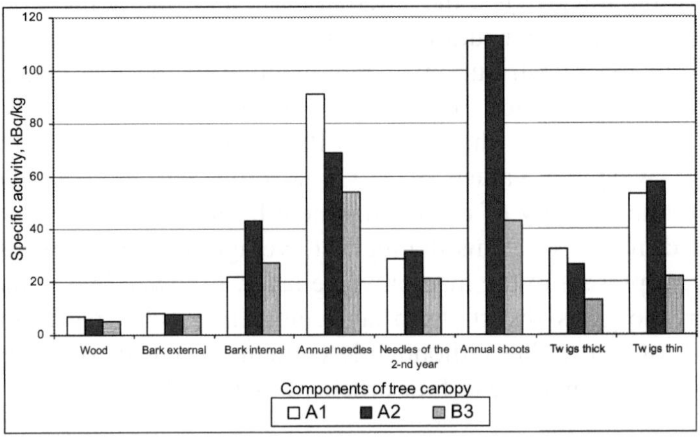

Figure 7. ^{137}Cs content in components of tree canopy in different forest cenosis.

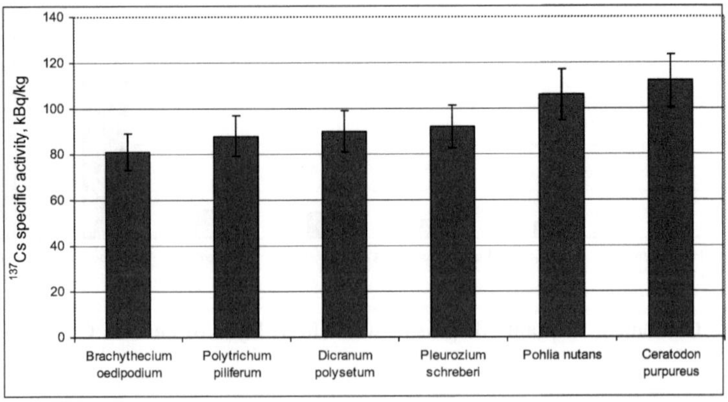

Figure 8. ^{137}Cs content in components of moss layer in cenosis Cladonio-Pinetum.

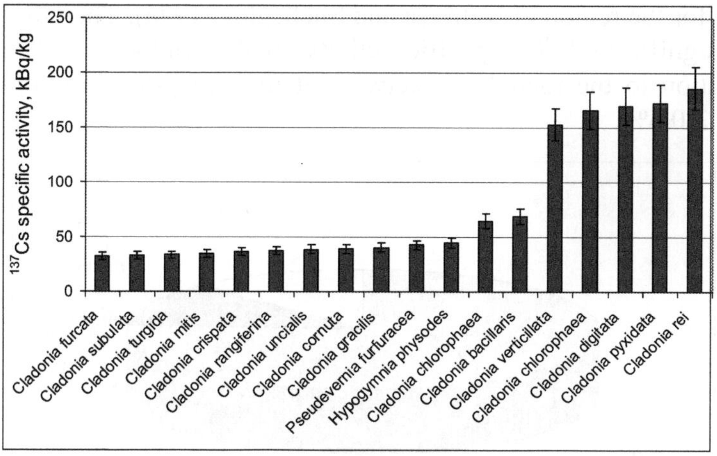

Figure 9. ^{137}Cs content in components of lichen layer in cenosis Cladonio-Pinetum.

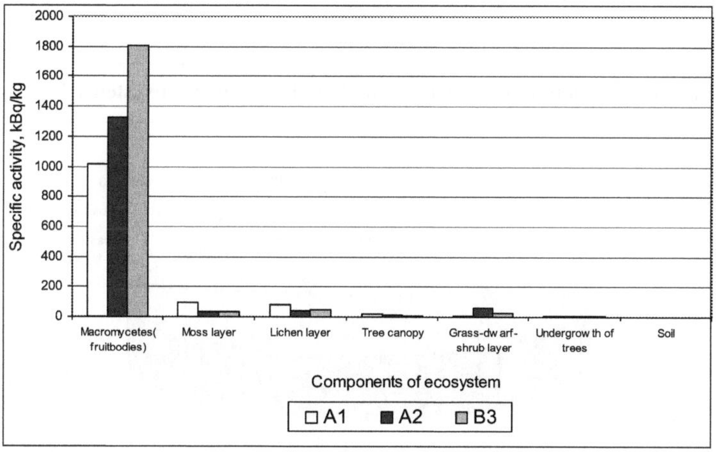

Figure 10. Average-weighted ^{137}Cs specific activity in different components of forest ecosystems.

In general, the total ^{137}Cs activity is non-uniformly distributed. In each forest ecosystem, three key components play a major role in the migration of ^{137}Cs (Fig. 11). Common to all ecosystems were the soil (including the forest litter and the mineral layers) and the tree canopy, containing about 88% of ^{137}Cs activity in the Cladonio-Pinetum, 90% in the Dicrano-Pinetum and 93% in the Molinio-Pinetum. Also important were the lichen layer (12% in the Cladonio-Pinetum), and the moss layer (9% in the Dicrano-Pinetum and 7% in the Molinio-Pinetum). Special attention was paid to the retention of ^{137}Cs activity by the mushroom complex. During the mushroom fruit body sampling period, the productivity of mushrooms on the experimental plots was low – about 6-8 times below the long term average for the region (Tsvetnova and Shcheglov, 1996). Therefore, in spite of the significant ^{137}Cs specific activity in the fruitbodies, their overall contribution to the total ^{137}Cs activity of the ecosystem was only in the range 0.2-0.6%.

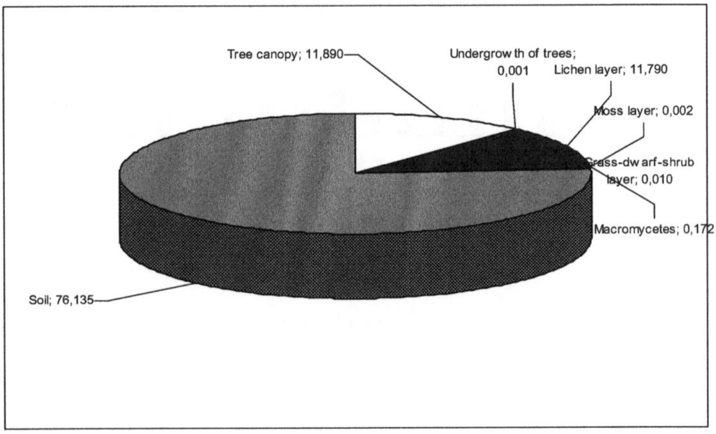

Figure 11a. The distribution of ^{137}Cs total stock in three forest ecosystems A_1.

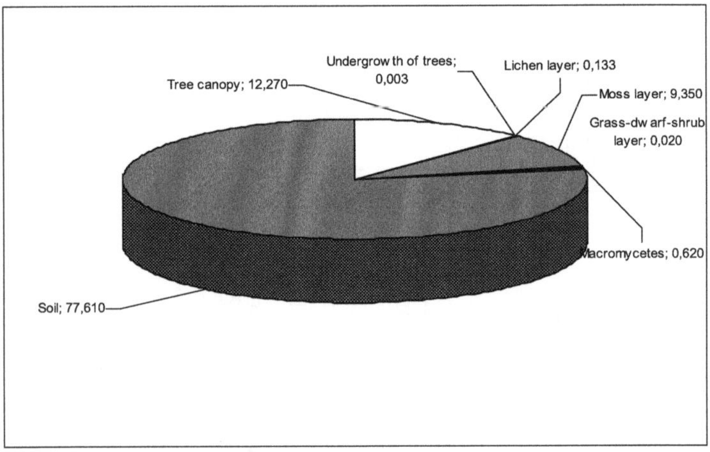

Figure 11b. The distribution of ^{137}Cs total stock in three forest ecosystems A_2.

Edible mushrooms are a significant element of the diet of the local population, and therefore represent a serious source of internal exposure to irradiation. Between 30% and 60% of the local population of Ukrainian Polissya regularly consume forest food products (Strand et al., 1996a), and a close correlation has been established between their consumption and ^{137}Cs load (Jacob and Likhtarev, 1996; Strand et al. 1996b). The contribution of forest food products to the internal exposure dose of the population in the forested region varies widely - from 12%-40% of the entire population to 50%-95% in some critical groups (Balonov et al., 1996;

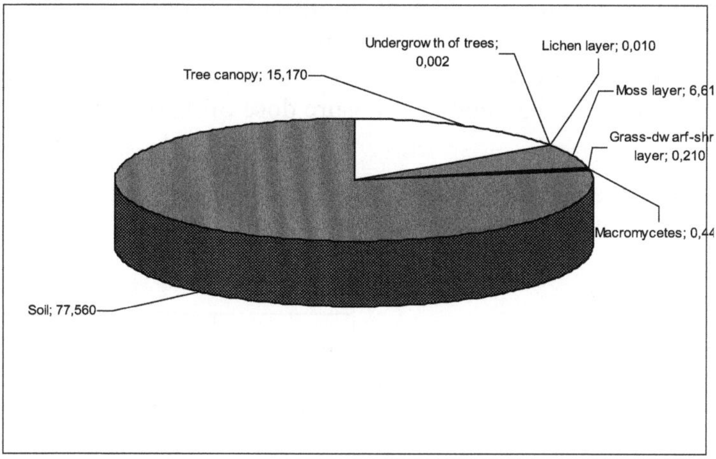

Figure 11c. The distribution of ^{137}Cs total stock in three forest ecosystems B_3.

Bruk et al., 1999; Giriy and Yaskovets, 1999). This variability is dependent on the density of ground radioactivity, the type of forest habitats, the species composition of wild mushrooms and berries, local peculiarities of the diet, and the culinary processing of food products (Kenigsberg and Belli, 1996; Kenigsberg and Buglova, 1994). For the present attempt to evaluate the contribution of forest food products to the internal exposure to gamma irradiation, we selected Khristinovka village, situated in the Narodichi district of the Zhytomyr region, in a heavily forested region. The village is in a region of high radio-contamination (^{137}Cs up to 100Ci/km^2) and although it is in the 'obligatory settling out zone', about 50 inhabitants remain. The surrounding pine forest is on sandy dunes, and there is no industry or agriculture. The population was divided into two groups according life style. Group A consisted of old, ill and single persons unable to support themselves financially. Their diet had a high proportion of 'forest gifts'. Group B, consisting of self-supporting young families with children, consuming a limited amount of 'forest gifts'. The average daily intake of each food product and its ^{137}Cs content were calculated. The extent of decontamination due to culinary processing was factored into the estimates for internal exposure dose (transfer coefficient of 14nSv/consumed Bq). Table 2 shows that the ^{137}Cs content in food products consumed by the villagers differed by 30,000 fold. The overall data describing the distribution of food products in the two groups and their contribution to internal exposure dose are presented in Fig. 12. For group A, agricultural (plant and animal) products contributed only about 2% of their total internal exposure dose, while 71% was derived from mushrooms, and 26% from bilberries. Over 94% of the dose was consumed from forest food products in the late summer and autumn, when the availability of forest mushrooms is highest. The average member of group A received, via their food, over 1MBq (14.65mSv) per year, while for those in group B, it was an order of magnitude less. Abstinence from forest food products would be expected to lead to a decrease in the internal exposure dose of 0.56mSv/year.

Table 2. Daily intake and ^{137}Cs content in food products of the population groups A and B in the village Khristinovka.

Food products	Daily intake, kg		^{137}Cs content, Bq/kg
	Group A	Group B	
Food products of plant origin from private house-economy			
Grain of wheat	0,29	0,26	6
Potato	0,35	0,32	24
Red beet	0,12	0,06	50
Onion	0,12	0,06	16
Sorrel	0,07	0,03	120
Carrot	0,05	0,09	65
Cucumber	0,4	0,05	5
Cabbage	0,27	0,24	3
Tomato	0,04	0,01	1
Haricot	0,06	0,04	30
Apples	0,2	0,2	7
Food products of animal origin from private house-economy			
Milk	0,8	0,3	100
Beef	0,055	0,05	290
Pork	0,05	0,04	200
Eggs	0,015	0,01	16
Food products of forest origin (“forest gifts”)			
Boletus edulis (dw)	0,0006	–	687000
Russula sp. (dw)	0,00005	–	2176360
Xerocomus chrysenteron (dw)	0,00005	–	6234120
Boletus edulis (fw)	0,006	–	45800
Cantharellus cibarius (fw)	0,003	–	85760
Russula sp. (fw)	0,0005	–	145090
Xerocomus chrysenteron (fw)	0,0005	–	415610
Berries of Vaccinium myrtillus (fw)	0,01	0,01	8420

Table 2. Continued.

Berries of Vaccinium myrtillus (dw)	0,008	0,001	84190
Meat of wild boar (fw)	0,003	–	1500
Meat of roe-deer (fw)	0,005	–	2000
Medicinal plants (average-weighted) (dw)	0,001	–	6000
River fish			
Pike (fw)	0,008	0,004	230
Perch (fw)	0,008	0,004	130
Roach (fw)	0,008	0,004	160
Crucian (carp) (fw)	0,008	0,004	100
Tench (fw)	0,008	0,004	60
Loach (fw)	0,008	0,004	60

*Note: density of ^{137}Cs ground deposition for fields and gardens – 678 100 kBq/m^2; for natural meadows – 1026 ± 250 kBq/m^2; forests – 1270 ± 350 kBq/m^2; river – 1026 ± 300 kBq/kBq/m^2; fw – fresh weight; dw – dry weight.

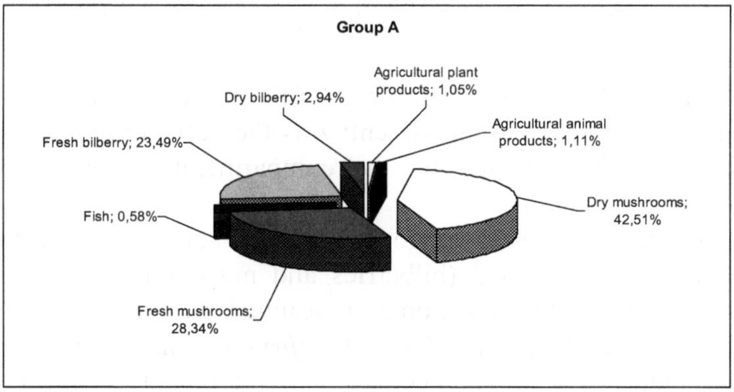

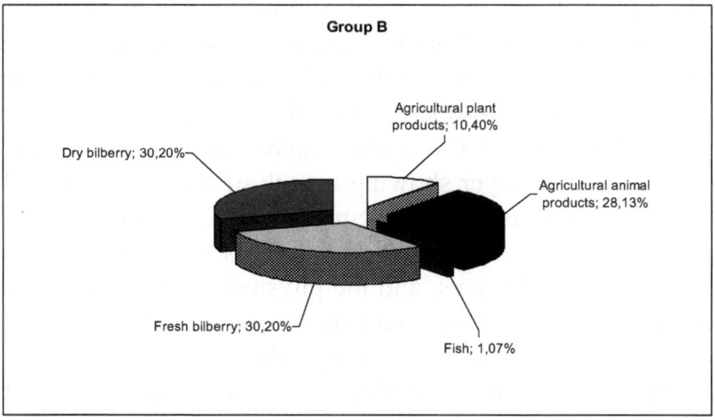

Figure 12. Contribution of different food products into internal exposure dose of population groups A and B of village Khristinovka.

5. Mathematical model history and description

The mathematical model for ^{137}Cs migration in the boreal forests of the Ukrainian Polissya was created on the basis of the ecosystem life-cycle, using the software system DRUADAN (Yanchuk et al., 2002).The boreal 55 year old forest ecosystem in the exclusion zone was the basis for assembly of the model, since long term radiological monitoring data of ^{137}Cs accumulation are available (Orlov et al., 2000). The anthropogenic influence on the ecosystem, in particular radionuclide fallout and its inclusion into the ecosystem life-cycle, was analysed (Strand et al., 1996). The model for the migration of ^{137}Cs was based on the compartmental method, as proposed by Georgievsky (1994), and the conceptual scheme used to create the model has been described in detail by Yanchuk et al. (2002). The structure of the model can be represented as a system of linear ordinary differential equations, where each equation describes ^{137}Cs migration into and out of the compartment:

$$\frac{dx_k}{dt} = \sum_{\substack{i=1,n,\\ i\neq k}} a_{i\_k} \cdot l_{i\_k} \cdot x_i - x_k \cdot \Big(\sum_{\substack{i=1,n,\\ i\neq k}} a_{k\_i} \cdot l_{k\_i} + \lambda + q_k \Big),$$

In this equations, l_{k_i} are the coefficients of the compartments' ties; x_i is the ^{137}Cs activity in a given compartment; λ is the half-life of ^{137}Cs; q is the coefficient of ^{137}Cs fixation by the compartment; a_{i_j} are the weight coefficients of the model.

On the basis of the average density of soil pollution, the ^{137}Cs accumulation in forest food (bilberries and mushrooms) was calculated, assuming that the edible mushrooms consumed were *B. edulis* (60%), *C. cibarius* (30%) and *R. paludosa, X. chrysenteron* (each 5%). The modelling indicates a strong decreasing trend (about 25 fold) of ^{137}Cs accumulation in O_l during the post-catastrophe period, particularly abruptly over the period 1986–1990. For O_f, there is a strong increase in the specific activity of ^{137}Cs during 1986–1990, followed by a slight decrease. The same dynamic behaviour applies to O_h,, but the peak of activity occurred during 1993–1994. The ^{137}Cs content increased for all the mineral soil layers, with the 0–2cm layer showing a peak of accumulation in 1997, the 2–4-cm layer in 1999, and the 4–6-cm layer in 2001. Following the peak, ^{137}Cs specific activity gradually decreased, as a combined result of its migration to deeper soil layers, and the physical decay of the radionuclide. Bilberries demonstrate a strong tendency to rapid self-decontamination in the initial period (1986–1992), during which the radionuclide content decreased three fold (and a further two fold from 1992-2002). This behaviour is mainly the result of the concentration of its root system in O_h and the 0–2cm mineral layer, where the availability of the radionuclide is high. The various edible mushroom species accumulate ^{137}Cs in a comparable way. The concentration of radionuclide increases for a certain period, then plateaus and final decays gradually. The modelling indicates that the maximal ^{137}Cs specific activity occurred in *C. cibarius* in 1997; in *X. chrysenteron* in 1996, in *R. paludosa* in 2000, and in *B. edulis* in 2005 (due to the greater depth of mycelium in this species). Measurement of the ^{137}Cs content of dry mushroom fruit bodies has shown that ten years after the Chernobyl catastrophe, the species rank as follows: *X. chrysenteron* >> *R. paludosa* > *C. cibarius* > *B. edulis*, and that after 30 years, *B. edulis* will have risen to the third position in this sequence.

A comparison of the mathematical model of the ^{137}Cs accumulation in forest food with actual radiological measurements made in 1999 validate the model. In particular, ^{137}Cs accumulation in materials gathered and prepared by the population of Khrystynivka were: dry *X. chrysenteron* 6.5± 2.0 MBq/kg, dry *B. edulis* 0.6±0.2MBq/kg; fresh *C. antharellus* 110±30kBq/kg; fresh bilberries 22±8kBq/kg. The ^{137}Cs accumulation in the major food products and forest food consumed by the local production was calculated using the mathematical modeling applied to the year 1996. The results are fully consistent with the measured amounts of exposure described above. A similar variability in radiation dose among the inhabitants of the forested regions of the Northern part of Ukrainian Polisya has been described by Orlov (1999), who measured a range of doses from 0.01 to 1.01mSv/year in one village, 0.01 to 2.07mSv/year in a second, and 0.03 to 1.9 mSv/year in a third. For modelling the internal exposure dose of the population of Khrystynivka in 2016 (i.e., 30 years after the Chernobyl catastrophe), it was assumed that the diet would remain constant, and that the availability of ^{137}Cs in the rhizosphere of agricultural land would not have significantly decreased. Decreasing the ^{137}Cs accumulation in food will then occur mainly due to radioactive decay and the migration down the soil profile of the radionuclides. The analysis indicated that the ^{137}Cs load of the Khrystynivka group A individuals will be significantly less than in 1996, reducing from 14.5 to 10.1mSv/year. However, the relative contribution to the dose from forest food will remain almost unchanged.

The dynamics of the impact of forest food load on the internal exposure dose of Khrystynivka groups A and B over the period 1996 and 2016 is of significant scientific and practical interest. For group A, the relative contribution of dry *B. edulis* to the internal exposure dose will have increased over two fold, that of *C. cibarius* will have fallen by 25%, that of *R. paludosa* will have slightly increased, and that of *X. chrysenteron* decreased. The relative contribution of fresh mushrooms will also have changed: *B. edulis* up about two fold, *C. cibarius* down by 1.3 fold, *R. paludosa* up by 1.1 fold, and *X. chrysenteron* up by 1.4 fold. The contribution of bilberries to the internal exposure dose of group A will have fallen by 3.5 fold. For group B, the contribution from forest food (mainly bilberries) will have decreased from 60% in 1996 to 33% in 2016, a 1.8 fold reduction.

6. Conclusion

For the inhabitants of the exclusion zone, situated in a forested region without well-developed economic and social infrastructures, it is of interest to estimate the relative contributions to internal exposure of forest food (berries and mushrooms) and conventional agricultural food (potatoes and milk). Some specialist studies have assumed that forest food represents only a small part of the diet of inhabitants of the exclusion zone, but we would suggest that in fact, it plays a major role. Typically, the consumption of forest food contributes 50% of the internal exposure dose, while for some critical population groups, it can exceed 80%. The definition of the spatial variation in pollution, the prevailing ecological conditions, and the diet should allow a scientifically-based prediction to be made of the internal exposure dose to the local population, and will help to identify the major risk factors within a certain time period following the Chernobyl catastrophe.

Acknowledgements

Special thanks are due to the NATO Science Programme Cooperative Science and Technology Sub-Programme who supplied support in the framework of the Collaborative Linkage Grant LST CLG. 980461 'Food Safety: risk analysis of living on the contaminated territories'. Our special thanks to www.smartenglish.co.uk for linguistic help in preparing this article.

References

Anuchin N.P., 1977. Forest taxation, Moscow, Lesnaya Promyshlennost Publ., 512 p.

Balonov M., Jacob P., Likhtarev I., Minenko V., 1996. Pathways, levels and trends of population exposures from consumption of agricultural and semi-natural products, The radiological consequences of the Chernobyl accident: Proc. of the 1-st Intern. Conference (Minsk, Belarus, 18-22 March, 1996), Luxembourg, pp. 235-251.

Bossew P., Lettner H., Hubmer A.K., 1996. Spatial variability of fall-out ^{137}Cs, in: Proc. of the Intern. Symp. on Radioecology 1996 "Ten years terrestrial radioecological research following the Chernobyl accident" (Vienna, 22-24 April, 1996), eds. M. Gerzabek, G. Desmet, B.J. Howard et al., Vienna, pp. 179-186.

Bruk G.Ya., Shutov V.N., Travnikova I.G., Balonov M.I., Kaduka M.V., Basalaeva L.N., 1999. The Role of the Forest Products in the Formation of Internal Exposure Dose to the Population of Russia after the Chernobyl Accident, in: Contaminated Forests. Recent developments in risk identification and future perspectives, eds. I.Linkov and W.R. Schell, Ser. 2, Environmental Security, **58,** Dordrecht–Boston–London, Kluwer Academic Publishers, pp. 343-352.

Dolin V.V., 2000. Criteria of self-cleaning of terrestrial ecosystems from radioactive contamination, Reports of NAS of Ukraine, **1:** 187-190.

Dolin V.V., Sobotovich E.V., 2000. Perspectives of natural-anthropogenic remediation of radioactively contaminated agroecosystems, Reports of NAS of Ukraine, **12:** 215-219.

Georgievsky V. Ecological and dose models in radiation accidents, 1994. – 236 p.

Giriy V., Yaskovets I., Kutlakhmedov Yu., Zaitov V. et al., 1999. Contribution of forest ecosystem to dose formation for inhabitans of the Ukrainian Polesye, in: Contaminated Forests. Recent developments in risk identification and future perspectives, eds. I. Linkov and W.R. Schell, Ser. 2, Environmental Security, **58,** Dordrecht–Boston–London, Kluwer Academic Publishers, pp. 325-332.

Jacob P. and Likhtarev I., 1996. Transfer factors for mushrooms, in: Pathway analysis and dose distributions, Final Report of JSP-5 for the contracts COSU-CT93-0053 and COSU-CT94-0091 of the European Commission, Luxembourg, pp. 63-74.

Kenigsberg J., Belli M., Tikhomirov F. et al., 1996. Exposures from consumption of forest products, The radiological consequences of the Chernobyl accident: Proc. of the 1-st Intern. Conf. (Minsk, Belarus, 18-22 March, 1996), Luxembourg, pp. 271-281.

Kenigsberg Y.E., Buglova E.E., 1994. Dose formation of internal exposure according to the peculiarities of casium radionuclides transfer by food chain and efficiency of countermeasures, in: Belarus-Japan Symposium “Acute and Late Consequences of Nuclear Catastrophes: Hirosima-Nagasaki and Chernobyl” (Minsk, Belarus, October 3-5, 1994), Proceedings, Minsk, pp. 82-97.

Mamikhin S.V., Tikhomirov F.A., and Scheglov A.I., 1997. Dynamics of Cs-137 in the forests of the 30-rm zone around the Chernobyl nuclear power plant, Sci. Total Environ., **193;** 169-177.

Melin J. and Wallberg L. Distribution and retention of cesium in Swedish boreal forest ecosystems, in: The Chernobyl fallout in Sweden, results from a research program on environmental radiology, ed. J. Moberg, Stockholm, Swedish Radiat. Protect. Inst., 1991, pp. 467-475.

Myakushko V.K., 1978. Pine forests of the plain part of Ukraine, Kiev, Naukova Dumka Publ., 256 p.

Orlov A.A., Krasnov V.P., Prishchepa A.L., 2002. Radioactive contaminated forests as a critical landscapes: radioactivity of food products and influence on forming of population internal exposure doses (analytical review), Zhytomyr, ZIET Publ., 202 p.

Orlov A.A., Turko V.N., Irklienko S.P., Krasnov V.P., Kalish A.B, Kurbet T.V., 1999. Use of forest resources on the base of forest-typological approach, VI Symposium on Radioecology (Szarvas, Hungary, 8-10.06.1999), **1**: Szarvas: 71-78.

Orlov O. Radiation situation and its dynamics in forest ecosystems of Polissya natural preserve in the period of quasi-equilibrium of the radionuclides in soil-plant layer covering (1991–1999 pp.) //30th anniversary of Polisya preserve. – Iss.1. – Zhytormyr, 1999. – Pp. 116–129.

Orlov O.O., Irklienko S.P., Krasnov V.P., Korotkova O.Z., 2000. Regularities of ^{137}Cs accumulation by wild mushrooms and berries in Polissya of Ukraine, Hygiene of Dwelling Places, **36(1)**, Kiev, pp. 431-445.

Orlov A., Irkliyenko S., Kalish O. Long-term dymanics of the radioactive pollution of boreal ecosystem compartments in Ukrainian Polissya // Proc. of “Radioactivity on nuclear explosions and accidents”. – Vol. 2.– St-Petersburg, 2000. – P. 249–254.

Pushkariov O.V., Davydchuk V.S., Sushchik Yu.Ya., Shramenko I.F., 2000. Evaluation of self-rehabilitation properties of environment, Reports of NAS of Ukraine, **2:** 208-213.

Shcheglov A.I., 1999. Biogeochemistry of technogenic radionuclides in forest ecosystems, Moscow, Nauka Publ., 268 p.

Shell W.R., Linkov I., Belenkaja E. Radiation dose from Chernobyl forests: Assessment using FORESTPATH model // Proc. of the 1-st international conference (Minsk, Belarus, 18-22 March, 1996). – Luxembourg, 1996.–P. 217–220.

Slavov V., Borschenko V. et all. Peculiarities of the internal and external exposure dose forming of the rural population of Exclusion zone // Visnyk of Agrarian Science. – April 2001. – P. 86–89.

Strand P., Balonov M., Skuterud L. et al., 1996. Exposure doses from consumption of agricultural and semi-natural products, The radiological consequences of the Chernobyl accident: Proc. of the 1-st Intern. conf. (Minsk, Belarus, 18-22 March, 1996), Luxembourg, pp. 261-269.

Strand P., Howard B., Averin V., 1996. Intake of radionuclides to man, in: Transfer of radionuclides to animals, their comparative importance under different agricultural ecosystems and appropriate countermeasures, Final Report of ECP-9 - EUR 16539EN, Luxembourg, pp. 157-193.

Tikhomirov F.A., Shcheglov A.I., 1997. Consequences of radioactive contamination of forests in affected zone of accident on ChNPP, Radiation biology. Radioecology, **37(4)**: 664-672.

Tsvetnova O.B., Shcheglov A.I., 1996. Accumulation of ^{137}Cs by higher macromycetes and their role in biogeochemical migration of nuclide in forest ecosystems, Review of Moscow State University, Ser. 17, Soil Science, **4**: 59-69.

Yanchuk V., Kolodnytsky M., Kovalchuk A., Levitsky V., Orlov O. Methods and tools of mathematical modeling of radionuclides migration in natural ecosystems: Vol 1, 2. – Zhytomyr: ZIET, 2002.

Yunatov A.A., 1964. Foundation of ecological profiles and experimental plots, in: Field geobotany, vol. 3, Moscow-Leningrad, Nauka Publ., pp. 9-35.

Zonn S.V., 1956. Interconnection and interdependence of forest vegetation with soils, Pochvovedenie, **7**: 80-90.

RISK ASSESSMENT AND REMEDIATION OF MILITARY AND AMMUNITION SITES

ANDRÉ GERTH, ANJA HEBNER
BioPlanta GmbH, Deutscher Platz 5, D-14103 Leipzig, Germany info@bioplanta-leipzig.de

*To whom correspondence should be addressed.

Abstract- Contamination of soil and water with explosives, especially 2,4,6-Trinitrotoluol (TNT), is a widespread problem on military sites and plants producing explosives and ammunition. Because the contaminated areas are often very large, off-site soil treatment technologies are prohibitively expensive (Gerth *et al.* 2003). Facilities for the treatment of explosive contaminated ground- and surface water have to be operated over decades. Since there is a need for cost-effective new technologies for the remediation of soil and water, biological technologies for the treatment of TNT in both media have been developed and applied. We present an overview of the *in situ* remediation of a former trickling pond by autochthonous micro-organisms, and the treatment of explosive-polluted surface water from a firing range.

Keywords: Military sites, explosives, TNT, *in situ* remediation, water treatment, constructed wetlands

1. Introduction

Military and ammunition sites contaminated with explosives can cover substantial areas (Gerth *et al.* 2005). Soil contamination in these sites is often heterogeneous. Explosives are relatively non-volatile, and have low aqueous solubility. Sampling from sites within a few decimetres of one another can result in concentration differences of up to one hundredfold (Jenkins *et al.* 1996). For example, the coefficients of variation across samples taken from 11 abandoned sites in the USA were 248% for TNT and 137% for Hexogen (Crockett *et al.* 1998). As a result, sampling error greatly exceeds measurement error. Thus to obtain representative results

N. Marmiroli et al. (eds.), Advanced Science and Technology for Biological Decontamination of Sites Affected by Chemical and Radiological Nuclear Agents, 45–57.

and a well based risk assessment, a large number of soil samples is necessary, especially on extensive sites.

Traditional sampling approaches use a large sampling raster, a small number of discrete samples and off-site analysis (RP-HPLC with UV or DAD-detection, GC-ECD). These methods are expensive. Because of the complexity of sample preparation, a long delay between sampling and outcomes is inevitable. High analytical costs and heavy time commitments can be reduced by using field analytical methods. The possibilities for on-site analysis of TNT are colorimetric methods, immunological methods, mobile GC´s and TNT sensors. An immunoassay for TNT (D-Tech®) from Strategic Diagnostics Inc., a Continuous Flow Immunosensor (FAST 2000/6000) from Research International and an Ion Mobility spectrometer are also available. The major colorimetric method is based on the Janovsky reaction, where acetone is added to the sample, and the NAC´s dissolve to form red coloured adducts (Jenkins 1990).

These methods correlate well with laboratory methods, as well as detecting the presence of other explosives (e.g. DNT´s) and need only little technical expertise. Using on-site analysis:

- many samples can be investigated within a short period of time
- contaminated sites can be rapidly identified and described
- information is provided regarding the heterogeneity of explosives in the soil
- subsequent laboratory analyses can be defined on the basis of the on-site analysis
- costs for laboratory analyses and risk assessment as a whole are reduced

Colorimetric tests have the drawbacks that they are not specific for a particular substance, and false positives can be drawn due to the presence of coloured humic contaminants. In the current project, we have used on-site colorimetric methods both to identify hot spots of contamination, and to select samples for laboratory analyses.

The remediation of contaminated soil with conventional technologies is very expensive. A viable option for these sites lies in a biological treatment technique. TNT and other explosives can be biologically transformed, immobilised and degraded by native micro-organisms, if they are stimulated by the addition of nutrients. A very effective agent for the bioremediation of explosives is molasses (Thomas *et al.* 2003). Overall, a reasonable strategy for risk assessment should include a field analytical method, while the processing of on-site materials can be The strategy for the processing of on-site materials should be based on detailed, extensive site investigations conducted by means of the raster technique, which will help to the identify areas affected by heavy contamination.

2. Remediation of a contaminated area on an ammunition plant

2.1. STARTING SITUATION

On the site of an ammunition plant, waste water contaminated with explosive was seeping into a trickling pond until 1982. The pond is about 300m long and was overgrown by trees and shrubs. For the site, a risk assessment was first undertaken. The soil in this area was highly contaminated with TNT and its transformation products. On-site concentrations of up to 50g/kg TNT were measured. A concentration of around 1g/kg was retained at 1m below the surface. Crystalline TNT (Figure 1) was identified in the soil in the vicinity of the pond inlet. Very high TNT concentrations were present in clay layers within the sandy soil. The whole contaminated area was about 2,000m². (Gerth *et al.* 2005) As a result of the risk assessment, a programme of soil remediation was considered necessary to protect the environment.

Figure 1. Crystalline TNT found on a contaminated site.

2.2. SELECTION OF REMEDIATION METHOD

On the basis of the risk assessment, various possible remediation technologies were considered. Excavation of the entire area and off-site incineration was an unacceptably expensive option. For the soil remediation of the former trickling pond, an innovative site-specific strategy was developed, involving three technological steps. First, soil with a concentration of above 1g/kg TNT (hot spots) was excavated to protect the groundwater. Second, a biological *in situ* remediation was initiated, following an on-site pilot test to estimate the remediation time and the TNT reduction which could be achieved by biological transformation. Third, a vegetative cover over the area was established to restore the site and to achieve long term protection of the groundwater. In comparison to full conventional treatment, the *in situ* approach is very favourable, both environmentally and financially. Excavation and off-site transport of only a small part of the contaminated soil were necessary.

2.3. SPECIFICATION OF THE TREATMENT STEPS

For the site of the trickling pond, a site-specific strategy for the remediation was established. Highly contaminated hot spots were excavated to protect the groundwater. The applied biological treatment technology was tested in pilot scale on site. A feasibility study was done for the estimation of remediation time and TNT reduction which can be achieved by biological transformation.

Finally a vegetative cover was planned for restoration and continuing groundwater protection.

2.3.1. *Excavation of hot spots*

Excavations were completed in summer 2002. The residues to be excavated were classified by on-site analytical methods, and were temporarily stored on-site for later off-site disposal. Soil sampling at the site showed that the highest level of contamination was found, as expected, in the area surrounding the former inlet of the pond. However, during excavation, a second hot spot was found with TNT concentrations of several g/kg TNT. Thus more soil had to be excavated than was initially expected. The TNT was concentrated in clayey soil.

2.3.2. *Treatment of soil by in situ bioremediation*

On the basis of the feasibility studies, an *in situ* technology was applied for the biological remediation of TNT and its transformation products. Microbiological transformation of TNT in the soil was stimulated by the addition of a carbon source and iron particles. The amount of additive was $5l/m^2$ molasses and $5kg/m^2$ iron particles. Thus for the treatment of the whole $2{,}000m^2$ site, there was a requirement for $10m^3$ molasses and 10t iron. The addition of molasses and a subsequent mechanical tillage to a

Figure 2. Excavation of the hot spots.

depth of 50cm was done quarterly, and while iron was added only once a year. The biological remediation in all took two years, with six treatmentepisodes. Treatment during the winter is ineffective because low soil temperatures allow little microbiological activity. The incorporation of additives was made affordable by the use of standard agricultural machinery (Figures 3 and 4). Following each soil treatment, the treated area was sown to grass and wild herbs.

During each treatment, 50 soil samples were taken, and monitored as follows:

1. For estimation of the heterogeneity of TNT contamination, five samples per cluster were taken, within a spacing of 50cm. The distance between adjacent clusters was 10m
2. All primary samples were analysed on site using photometric tests
3. The five primary samples per cluster were combined into a single composite sample
4. Composite samples were analysed by GC/ECD

Figure 3. Addition of molasses.

Figure 4. Mechanical tilling of the soil.

The decrease in TNT concentration during the *in situ* treatment is illustrated in Figure 5. TNT concentrations up to 1g and 0.1g/kg measured in at least one primary sample per cluster are included in this figure. TNT concentration was significantly reduced by the *in situ* remediation, and the target reductions were attained. The average TNT concentration before the first *in situ* treatment was 372mg/kg, but after the sixth treatment, it had fallen to about 23mg/kg.

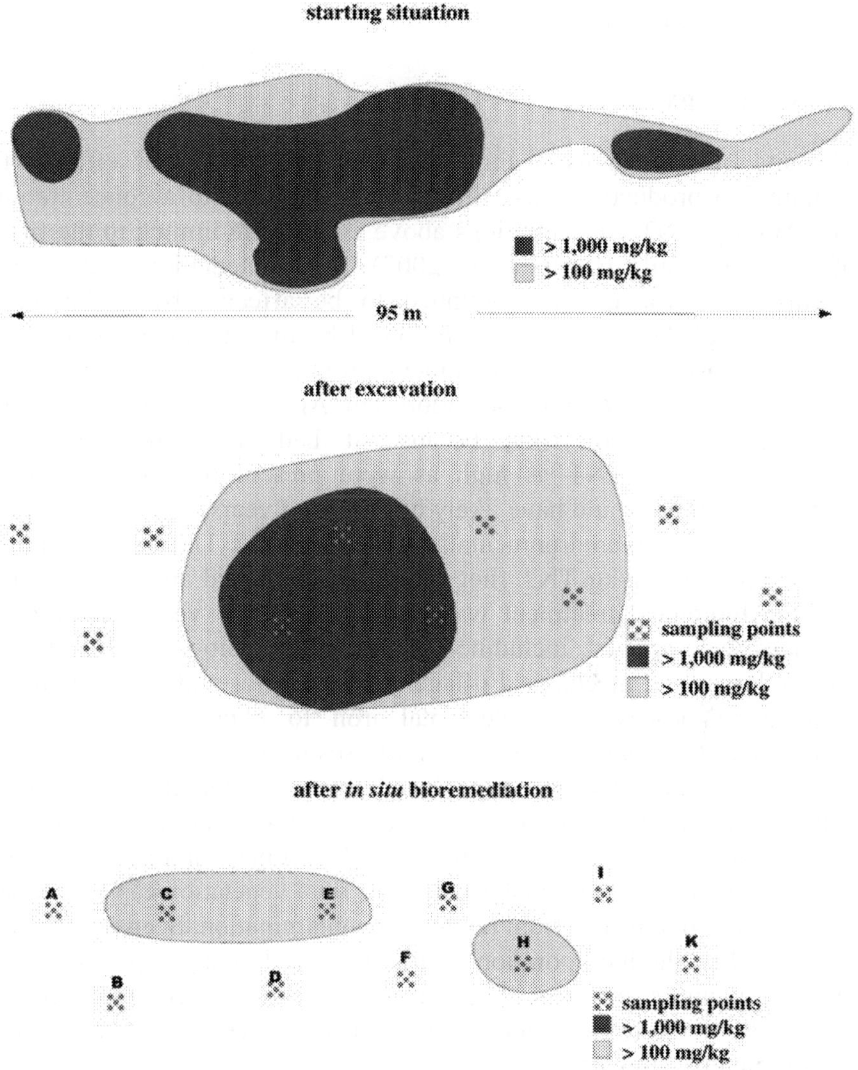

Figure 5. TNT concentration in the soil.

2.3.3. *Vegetative covering*

In the final step, the site of the former pond was capped first with a layer of soil with a high water storage capacity (clay) followed by fresh top soil brought from elsewhere. Grass was sown, and poplar and willow trees planted on the site, since established vegetation minimizes the formation of leachate, by generating a flow of evapotranspiration. The combination of grass and trees was chosen to develop the vegetative cover in the shortest possible period of time.

2.4. SUMMARY

The site of a pond used until 1982 for the disposal of effluent from ammunition production was polluted with TNT, up to a concentration of over 50g/kg. TNT concentrations above 1g/kg were limited to the first 1m of the soil profile (Gerth et al. 2005). Soil remediation was necessary because the groundwater continued to be affected by pollution. The excavation of the complete area followed by off-site incineration was too expensive, so a multi-stage treatment programme was developed which incorporated an *in situ* bioremediation step. An *in situ* bioremediation alone was considered too risky, because it had never been tried with concentrations of TNT as high as were present on this site, and the remediation time would have likely been tens of years. The *in situ* approach is very favourable, environmentally and financially. The soil contaminated with more than 1g/kg TNT (hot spots) was removed and treated off-site, and the biological treatment was realised into two years. Six treatment episodes were applied, including the combined addition of elemental iron and molasses and mechanical tillage. Molasses was added to stimulate the degradation process, and elemental iron to obtain strongly reducing conditions. Cyclical addition of molasses/iron and mechanical tillage resulted in an even distribution of the additives, good homogenisation of the soil and the cycling of anoxic and oxic conditions. The process affects the uppermost 50cm of the soil profile. The TNT concentration was reduced to 0.1mg/kg by biological transformation. A final vegetation cap was added to protect the groundwater from remaining contamination. Seepage of water is minimised by the incorporation of a high water holding capacity soil layer and the planting of trees. The long-term phytoextraction of remaining pollutants by the trees should allow most of the residual contamination to be completely removed.

3. Biological treatment of surface water in a wetland system contaminated with explosive

3.1. STARTING SITUATION

On a military training area in northern Bavaria (Germany), four million soldiers have been trained since 1936. A part of this training area is still actively used as a firing range to test explosives. On average, 1 to 2% of the ordinance is not exploded, and the explosives are leached out by rain. As a consequence of continual detonation activities, the underground (sand stone) had become compacted and porous. These changes in the geological formation of the soil have led to an increase in the quantity of leachate of unexploded ordinance. Because of the loamy top soil, rainwater puddles at the surface. Only limited use was still being made of the detonation area for military activities was limited. In a pilot project, the actively used detonation ground was reconstructed and an artificial wetland was for the treatment of drainage water was planned and built.

3.2. RECONSTRUCTION OF THE DETONATION GROUND

The top layer of the detonation ground was excavated to a depth of 0.5m. The excavated soil was temporarily stored on the site. A Bentonit seal was installed and a protective and a drainage layer were built over the seal. The excavated soil was incorporated into the slopes of the reconstructed ground (Figure 6). The new top layer of the detonation ground was made from

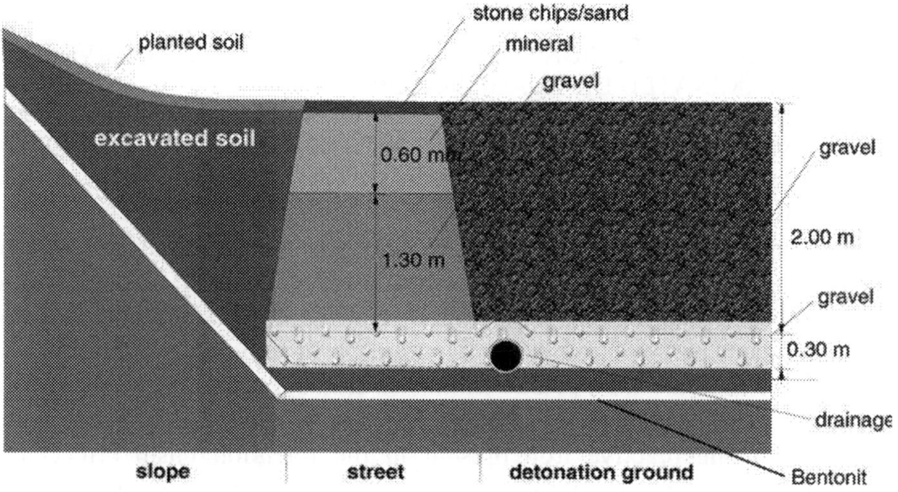

Figure 6. Scheme of the design from the reconstructed detonation ground.

Figure 7. Laying of the Bentonit mats.

coarse-grained gravel. Compressed stone chips were built into the road surface. The levelling of detonation craters on the reconstructed detonation site was necessary to protect the Bentonit seal. The drainage water flows through several shafts and a grit chamber into the constructed treatment plant.

3.3. TECHNICAL SOLUTION FOR THE TREATMENT OF EXPLOSIVES CONTAMINATED DRAINAGE WATER

A retention basin and a constructed wetland were planned and built for the treatment of drainage water polluted with TNT, Hexogen and Octogen. The maximum concentrations of explosives measured in the surface water in 2000 were 7.4mg/l TNT, 0.37mg/l Hexogen and 0.055mg/l Octogen (Emmert *et al.* 2004). The concentration of explosives is determined by the amounts of precipitation experienced and explosives detonated. The drainage water passes through the retention basin into the wetland, both of which were sealed with a plastic film. The drainage water was collected in the retention basin. The concentration of explosives in the drainage water was stable. The reactive iron present in the retention basin effects the removal of oxygen, while the gravel wall inside the basin retains solid substances. The pre-treated and oxygen-reduced water then flows into a shaft, and thence, at a defined flow rate, into the constructed wetland. For more effective microbiological transformation of the explosives, molasses were added to the water. After passing the constructed wetland, the treated water was then discharged into a water trench. Sensors for the measurement of nitrate, pH and water temperature were installed in the outflow shaft from the wetland. The measurement of these parameters in the effluent is necessary to regulate the dosage of molasses. The automated data collection enabled rapid changes to be made to the flow rate and/or the dosage of molasses by changing the redox conditions prevailing in the wetland. The energy for these facilities was generated by a solar panel.

Since the beginning of its operation, several technical problems have affected the treatment plant. An intensive test phase was required for its optimization. The reduction of 2-ADNT, 4-ADNT, which are the transformation products of TNT, is critically affected by changes in the air temperature during the autumn and spring. In these seasons, the firing range was intensively used and high rainfall is normal at the site. Prescribed limits for explosive compounds do not presently exist in Germany, but the responsible public authority had defined target values. These target values were reached during the failure-free operation of the plant.

Figure 8. After excavation of soil and make the profile placement of the folia.

Figure 9. Placement of the inflow and outflow drainage, filling of the gravel.

Figure 10. Treatment plant after plantation and start of operation.

3.4. SUMMARY

On an operating firing range used by the German army, residual explosives are eluted by rainfall. The detonation area was reconstructed including a drainage system. A treatment system was built for the remediation of drainage water contaminated with the explosives TNT, Hexogen and Octogen. The drainage water was collected and fed into a retention basin and an artificial wetland. For the microbial transformation of explosives, reducing conditions are needed. Hence the retention basin was designed to contain reactive iron. Molasses was added via the inlet to the wetland. The water treatment system is very sensitive to changes in the prevailing redox conditions. The limits for the COD, BOD, total nitrogen and total phosphate can be reliably reached. Following site-specific optimisation of the operation parameters during the test phase, the target values for the explosive compounds were met. The reconstruction of the detonation ground and the construction of the treatment plant were achieved using Army personnel, and the water treatment system needs only little maintenance and low operating costs. We are currently planning to channel the surface water from another firing range on the same military site into the treatment plant.

Acknowledgements

This work was partially supported by Mr. Witt (Rheinmetall DeTec AG, Germany), Mr. Emmert (Standortverwaltung Hammelburg, Germany), and funded by BMBF under grant Nr. 02 WU 0101, 02 WU 0102).

References

Crockett A. B., Jenkins T. F., Craig, H. D., and Sisk, W. E., 1998, Overview of On-Site Analytical Methods for Explosives in Soil, U.S. Army Cold Regions & Engineering Laboratory, Special Report 98-4.

Emmert, H., Kuhne, A., and Thomas, H., 2004, Remediation and environmentally sustainable reconstruction of a Ordnance Detonation Site, Mutual Weapons Development Master Data Exchange Agreement between the Government of the united States and the Government of the Federal Republic of Germany, 14th General meeting 2004, Berlin, 16. – 20. February 2004.

Gerth, A., Hebner, A., and Thomas, H., 2003, Case Study Natural Remediation of TNT-Contaminated Water and Soil, *Acta Biotechnol.* 23:143-150.

Gerth, A., Thomas, H., and Hebner, A., 2005, Biologische In-situ-Sanierung von TNT kontaminiertem Boden, TerraTech 1-2, TT 21-23.

Jenkins, T. F., 1990, Development of a Simplified Field Method for the Determination of TNT in Soil. U.S. Army Cold Regions Research & Engineering Laboratory, Special Report 90-38.

Jenkins, T. F. e.a., 1996, Assessment of Sampling Error Associated with Collection and Analysis of Soil Samples at Explosives-Contaminated Sites. U.S. Army Corps of Engineers, Cold Regions Research & Engineering Laboratory, Special Report 96-15.

Thomas, H., Hebner, A., and Gerth, A., 2003, Nutzung des biologischen Abbaus von TNT zur Sanierung von Rüstungsaltlasten, GWF Wasser Abwasser, 144, Nr. 5, 355-358.

PART 2

BIOLOGICAL BASES OF BIOLOGICAL REMEDIATION

MICROBIAL SYSTEMS FOR IN-SITU SOIL AND GROUNDWATER REMEDIATION

L. DIELS*, R. LOOKMAN
Environment and Process Technology, Flemish Institute for Technological Research (VITO), Mol, Belgium,
ludo.diels@vito.be

*To whom correspondence should be addressed.

Abstract- The contamination of groundwater by leakage of hydrocarbons or other pollutants from underground storage tanks, distribution systems and various industrial operations is a major environmental problem. Conventional treatment techniques suffer from serious shortcomings which limit their applicability and efficiency. These include high cost and maintenance requirements, the need to transfer the contamination from one medium to another, and the extended duration of the operation. An alternative to these treatments lies in *in situ* remediation. Biochemical reactions play an important role in many of these *in situ* remediation technologies. Commonly, micro-organisms degrade organic compounds to recover the chemical energy conserved in the C-C bonds. Complex organic molecules are thereby converted to simpler compounds and, ultimately, to carbon dioxide or methane and water. Parts of some compounds are scavenged to provide the building blocks for new bacterial cells. *In situ* bioremediation relies on the ability of micro-organisms (bacteria, fungi, algae, protozoa and metazoa), either attached to soil particles or suspended in the pore water, to either convert contaminants into harmless molecules such as carbon dioxide and water, or to immobilize them into non-soluble and stable forms. In *natural attenuation* micro-organisms just present in the contaminated aquifer degrade o precipitate the present contaminants by consuming naturally present electron donors or acceptors. However, in most cases electron acceptors are not in sufficient supply, and are rapidly exhausted. In this case, biodegradation may be induced by the injection of an appropriate electron acceptor. For highly oxidized pollutants particularly, electron donors need to be injected. The injection of an electron acceptor or donor creates a subsurface zone where migrating

N. Marmiroli et al. (eds.), Advanced Science and Technology for Biological Decontamination of Sites Affected by Chemical and Radiological Nuclear Agents, 61–77.

contaminants are intercepted and permanently immobilized (for instance via the bioprecipitation of for heavy metals) or converted to harmless end-products. This process is called *biostimulation.* In certain cases, bacteria able to degrade the pollutants present are absent, or present in insufficient density in the natural soil environment. In this case extra specific bacteria need to be injected into the aquifer. This process is called *bioaugmentation.* The selection of the appropriate remedial technology at a particular site will be determined by time constraints for attaining the remediation objectives, the site's hydrogeology and geochemistry, the particular contaminants involved, prevailing regulatory constraints, and considerations of environmental exposure and cost. Even though environmental disturbances can be modified by micro-organisms, microbial ecosystems lack long-term stability and are continually adapting. In designing a fail-safe bioremediation system, it is therefore important to understand the complexities and interactions within the target ecosystem. This paper will deal with these basic understandings.

Keywords: Micro-organisms, biodegradation, bioprecipitation, biostimulation, bioaugmentation, electron donor, electron acceptor, injection.

1. Introduction

The contamination of groundwater by leakage of hydrocarbons or other pollutants from underground storage tanks, distribution systems and various industrial operations is a major environmental problem. Conventional treatment techniques suffer from serious shortcomings which limit their applicability and efficiency. These include high cost and maintenance requirements, the need to transfer the contamination from one medium to another, and the extended duration of the operation, since decades may be necessary to prevent continued growth of contaminant plumes (Yerushalmi et al., 1999). An alternative to these treatments lies in *in situ* remediation.

Biochemical reactions play an important role in many *in situ* remediation technologies. Bioremediation approaches are characterised by considerably lower cost and maintenance than are typical for *ex situ* technologies. Contaminants are destroyed or immobilized on site, avoiding the need for their transport to the surface for on- or off-site treatment. As a result, energy is saved, and the risk of undesirable exposure or migration of contaminants is reduced. *In situ* bioremediation relies on the ability of micro-organisms (bacteria, fungi, algae, protozoa and metazoa), either attached to soil particles or suspended in the pore water, to either convert contaminants into harmless molecules such as carbon dioxide and water, or to immobilize them into non-soluble and stable forms. It can be usefully combined with certain *in situ* chemical or physical techniques; for example, zerovalent iron used in permeable reactive barriers against chlorinated hydrocarbons, produces hydrogen gas that serves as an effective electron donor for a number of microbial reactions. A schematic representation of the process of microbial degradation in a soil pore is presented in Figure 1. Contaminants attached to soil particles (gravel, sand, silt, clay, organic matter) first diffuse into the pore water, and enter the microbial cell via specific uptake mechanisms. Bacteria use an electron acceptor (typically oxygen) to convert, using very specific enzymes as catalysts, the pollutant into biomass and harmless products. In the process they consume those nutrients (particularly nitrogen and phosphorus) necessary for the generation of biomass. Some fungi excrete enzymes that chemically modify pollutants present in their vicinity.

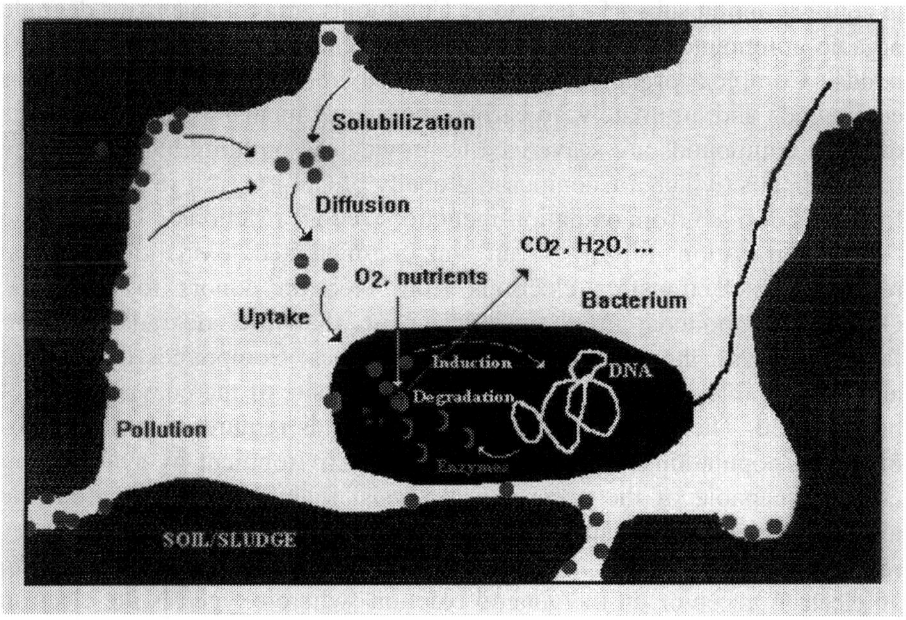

Figure 1. Schematic presentation of microbial degradation of pollutants (red dots) in a soil pore (Leen Bastiaens, personal communication).

Microbial communities are in a state of flux and are constantly adapting to their environment. Population dynamics, environmental conditions, and growth substrates are subject to continuous change, and this results in complex interactions between microbial populations. Even though environmental disturbances can be modified by micro-organisms, microbial ecosystems lack long-term stability and are continually adapting. In designing a fail-safe bioremediation system, it is therefore important to understand the complexities and interactions within the target ecosystem.

2. Microbial processes

Micro-organisms have evolved a variety of characteristics that allow them to survive and distribute themselves in the environment. Bacteria are by far the most prevalent and diverse of the micro-organisms. Fungi live predominantly in soil or on dead plant material, and are responsible for the mineralization of organic carbon and the decomposition of cellulose and lignin. Typically, micro-organism density decreases with increasing depth in the soil profile, as does the content of organic matter. However the population does not decrease to extinction with increasing depth, nor does it necessarily reach a constant density. Micro-organisms are more numerous in silty or silty-clay soils than in sandy or coarse sandy matrices. Most fungi prefer the upper soil profile. The rhizosphere (root zone) contains the greatest variety and number of micro-organisms. Microbial life also thrives in aquifers and in subsurface regions. Commonly, micro-organisms degrade organic compounds to recover the chemical energy conserved in the C-C bonds. Complex organic molecules are thereby converted to simpler compounds and, ultimately, to carbon dioxide or methane and water. Parts of some compounds are scavenged to provide the building blocks for new bacterial cells (which are composed globally of $CH_{1.8}O_{0.5}N_{0.2}S_{0.0046}P_{0.0054}$). Energy is derived from oxidation-reduction (redox) reactions.

Biodegradation relies on an enzyme-mediated oxidation/reduction reaction which transfers electrons from electron donors to acceptors. Organic compounds in the environment that are degradable align favourably with the active site of such enzymes. Compounds that do not align favourably or do not bind with the active site of the enzyme will not be degraded. The breakdown of such compounds requires that the micro-organism population adapts to an unfamiliar environment by synthesizing enzymes capable of the necessary catalysis. In biological oxidation, the target is degraded by a catalytic pathway which thereby provides the energy required for cell metabolism and synthesis. The target donates electron(s) to an electron acceptor, in the general reaction (where oxygen is the electron acceptor):

$$CH_{1.8} + \alpha O_2 + \varepsilon NH_3 \rightarrow \beta CO_2 + \gamma CH_{1.8}O_{0.5}N_{0.2} + \delta H_2O$$

In addition to carbon, nitrogen and phosphorus are also essential to produce biomass. Other micro-nutrients such as sulphur, potassium, sodium, calcium, magnesium and some metals are necessary to maintain microbial viability.

The yield (γ) of a biomass production process is defined as the moles of biomass formed per mole of substrate consumed. Aerobic conditions are more conducive to higher biomass formation (and therefore also to biofilm formation) than anaerobic conditions. Empirically, under aerobic conditions, a yield of 0.05 – 0.6mol biomass/mol carbon can be obtained, while under anaerobic conditions the attainable yield falls to 0.04 – 0.083mol. The reaction kinetics of biodegradation processes can be approximated by the first-order reaction rate constant k as follows:

$$C = C_0\, e^{-kt} \text{ or } \ln C/C_0 = -kt$$

Thus the half-life can be calculated as: $t_{1/2} = 0.693/k$

An overview of typical reaction rate constants found in literature is given in Table 1. The large range in reported rates prevents these rate constants being extrapolated to any particular biodegradation situation, so that only a site-specific study can yield reliable kinetic information. Parameters or conditions influencing biodegradation rates are:

- Humidity
- Temperature (5 – 30°C)
- pH : 5-8
- Toxicity (i.e. other pollutants)
- Pollutant distribution and bio-availability:
    - as solid particles
    - low solubility
    - adsorbed to the soil matrix
    - bound to organic matter
    - via diffusion in micro-pores
    - threshold concentration

Micro-organisms can produce energy via three independent processes: (1) aerobic — requiring oxygen; (2) anaerobic — relying on nitrate, iron, manganese, sulphate or carbonate; and (3) fermentation — relying on organic compounds to act both as electron donor and acceptor. In principle, in any environment in which microbial activity occurs, there is a gradient from aerobic to anaerobic/methanogenic conditions. A defined sequence of electron acceptors is used in this progression through distinct redox states. The following scenario outlines a general sequence of events in which the oxidation of a carbon source occurs first. The carbon source may be the

Table 1. Some degradation rate constants and half-life times taken from literature.

	Slow degradation		Fast degradation	
	k	$t_{1/2}$	k	$t_{1/2}$
PCE	0.00019	3547	0.0033	210
TCE	0.00014	4950	0.0025	277
VC	0.00033	2100	0.0072	96
TCA	0.013	533	0.01	69
Benzene	0	-	0.0036	193
Toluene	0.00099	700	0.015	12
Ethylbenzene	0.0006	1155	0.015	46
m-xylene	0.0012	578	0.016	43
o-xylene	0.00082	845	0.021	33
p-xylene	0.00085	815	0.015	46

contaminant(s) of interest or other, more readily degradable organic compounds present. Once all the available oxygen is consumed, active aerobic populations begin to shift towards nitrate respiration. This denitrification process will continue until all the available nitrate is depleted, or until the carbon source becomes limiting. Thereafter, manganese-reducing populations take over. Iron reduction does not occur until all available manganese IV oxides are reduced. In addition, bacterial Mn(IV) respiration appears to be restricted to where sulphate is almost or totally absent. At lower redox conditions, sulphate-reducing bacteria dominate and once carbon source or sulphate limitations apply, methanogenic bacteria begin to dominate. An overview of the energy gain and electron acceptor use for the degradation of toluene is presented in Table 2. Overall, the use of oxygen as the electron acceptor is the most energetically favourable reaction, and therefore the introduction of oxygen into soil and groundwater is likely to stimulate a number of biodegradation reactions. One litre of water (saturated with oxygen) contains about 8mg oxygen, while one litre of air contains about 300mg. Thus, per litre pore

Table 2. Oxidative toluene degradation in the presence of electron acceptors.

Electron donor	Electron acceptor	Reaction products	ΔG (kJoule)	ORP (mV)
$C_6H_5CH_3$	O_2	$CO_2 + H_2O$	3875	+ 820
	NO_3^-	$CO_2 + H_2O + N_2$	3823	+ 740
	MnO_2	$CO_2 + H_2O + Mn^{2+}$	3824	+ 520
	FeOOH	$CO_2 + H_2O + Fe^{2+}$	2792	- 50
	SO_4^{2-}	$CO_2 + H_2O + H_2S$	598	- 220
	CO_2	$CO_2 + H_2O + CH_4$	143	- 240

volume, 37.5 times more oxygen is present in unsaturated , as compared to saturated soils.

3. Natural attenuation

Interest in the natural attenuation of groundwater contaminants has increased in recent years, given the complexities of subsurface systems and the inherent problems and costs associated with conventional remedial technologies, such as pump-and-treat systems. The OSWER (Office of Solid Waste and Emergency Response) of the US-EPA has defined natural concentration of pollutants in these media". While the mechanisms controlling the chemical transformation, dispersion, dilution, sorption and volatilization processes have been identified, aerobic and anaerobic degradation remain the major processes for the reduction of contaminant material in the subsurface. The use of these naturally occurring processes provides an option for the restoration of contaminated sites. The US-EPA term 'monitored natural attenuation' refers to the exploitation of natural attenuation to achieve site-specific remedial objectives within a time-frame that is reasonable compared to more active methods. In Figure 2, an overview is provided of the various electron acceptor zones in a contaminated groundwater plume.

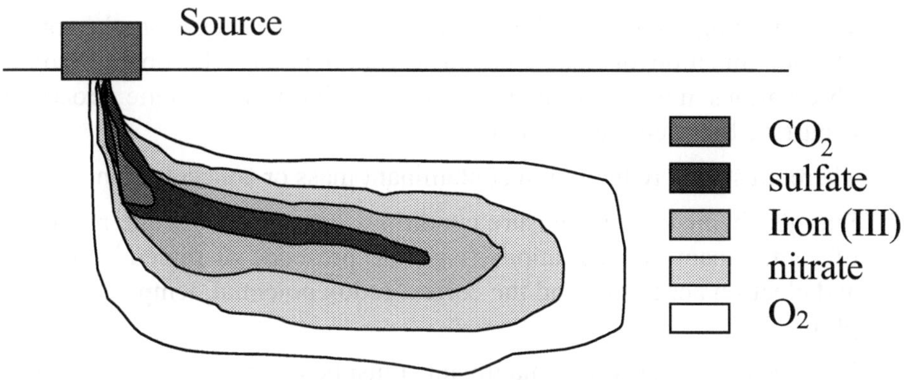

Figure 2. Biodegradation in a groundwater plume causing different redox zones.

As a result of the highly reduced state of petroleum hydrocarbons, the preferred and most thermodynamically terminal electron acceptor for microbial processes is oxygen. The inverse relationship between the concentrations of BTEX and dissolved oxygen within a plume is indicative of the extent of microbial metabolism of this class of contaminant. Data from various sites indicate that the natural attenuation of BTEX proceeds at higher rates under oxygenated conditions. The biodegradation of chlorinated solvents, depending on the degree of halogenation, is fundamentally different from that of petroleum hydrocarbons and other oxidized chemicals, and the preferred redox condition is anaerobic. The terminal electron acceptor process that predominates in certain parts of the contaminant plume is reflected by the local hydrogen concentration. Hydrogen concentrations for various terminal electron acceptors are shown in Table 3.

Table 3. Terminal electron accepting processes and typical related hydrogen concentrations.

Terminal electron accepting process	Hydrogen (H_2) concentration (nmol/l)
Denitrification	< 0.1
Iron (III) reduction	0.2 – 0.8
Sulphate reduction	1 – 4
methanogenesis	5 - 20

The behaviour of a contaminant plume, whether stable, shrinking or expanding, is predictive for the occurrence of natural attenuation. However, few historical data describing the status of a plume are available. Thus to confirm that natural attenuation processes are taking place, at least four basic conditions must pertain (Azadpour-Keeley et al. 1999), specifically:

5. The sampling points must be on groundwater flow lines from the source of contamination or above the observation point. The down-gradient observations must accurately reflect the abiotic and biotic processes occurring between the two points.
6. There must be a reduction in contaminant mass or concentration.
7. Site geochemistry must ensure conditions appropriate for the reduction of contaminant concentrations (e.g., the presence of mineral nutrients and electron acceptors, and the correct redox potential, temperature, and pH).
8. Breakdown products of contaminants must be present, with indicators of mineralization. In addition to the biological utilization of oxygen, nitrate, and sulphate, these natural attenuation processes often result in the appearance of dissolved Fe(II), Mn(II), HCO_3^-, CO_3^{2-}, and methane. Hendrickx et al. (2005) have described certain *in-situ* microcosm systems designed to track natural attenuation processes.

4. Biostimulation

As mentioned above, soil and groundwater contain a wide variety of micro-organisms able to cope with different pollutants. In some cases natural attenuation is effected by autochthonous microflora using various available electron acceptors. However, in most cases electron acceptors are not in sufficient supply, and are rapidly exhausted. In this case, biodegradation or bioprecipitation (for heavy metals) may be induced by the injection of an appropriate electron acceptor. For highly oxidized pollutants particularly, electron donors need to be injected. The injection of an electron acceptor or donor creates a subsurface zone where migrating contaminants are intercepted and permanently immobilized or converted to harmless end-products. The successful design of such a 'reactive zone' requires the establishment of two sets of reactions: (1) between the injected reagents and the migrating contaminants, and (2) between the injected reagents and the subsurface environment. An alternative to a 'reactive zone' is a 'permeable reactive barrier', which is typically installed by creating a trench perpendicular to the contaminant plume, and filling it with carrier material to provide the necessary nutrients and micro-organisms (Figure 3). A range of processes can be induced inside the barrier boundaries by the injection of electron acceptors and/or donors.

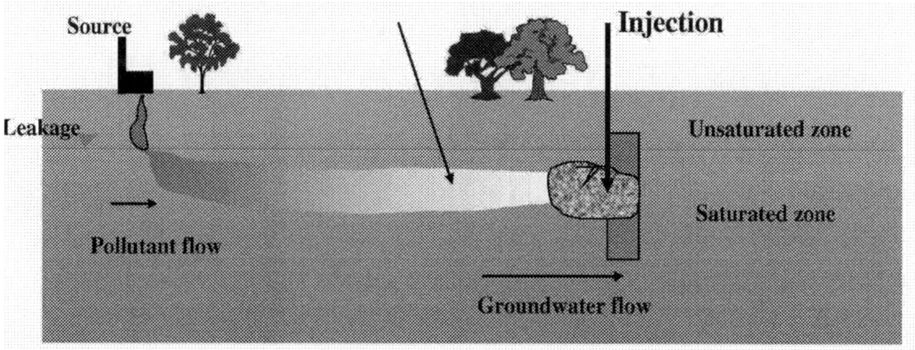

Figure 3. Reactive Zone or Permeable Reactive Barrier to treat a contaminant plume *in- situ*.

4.1. ELECTRON ACCEPTOR REACTIONS

Oxygen is the most important electron acceptor used by micro-organisms, and in many cases biodegradation can be stimulated by the addition of oxygen to the soil or groundwater. In so-called biosparging, air is introduced into the groundwater via sparging wells (Figure 4). This process results in the stripping from the groundwater of volatile hydrocarbons such as chlorinated aliphatics and BTEX, after which they can be isolated by soil vapour extraction. At the same time, oxygen is introduced at a maximum concentration of 10 mg/l at 12 °C. If pure oxygen is injected, for example via an ISOC (*in-situ* oxygen curtain) system, about 50 mg/l can be achieved. Other potential oxygen sources that can be introduced in the soil or aquifer are ozone (which decomposes into oxygen) and hydrogen peroxide (generating oxygen and water). Solid oxygen-releasing compounds (e.g. MgO_2 powder [ORC® or Oxygen Release Compounds]) are also available, and these can be injected or introduced via socks in vertical wells. The oxygen is slowly released via the reaction:

$$MgO_2 + H_20 \rightarrow \frac{1}{2}\, O_2 + Mg(OH)_2$$

Oxygen introduction is appropriate for the degradation of BTEX, mineral oil, phenols, MTBE, PAHs and vinyl chloride. In rare cases, electron acceptors such as nitrate of sulphate may be considered.

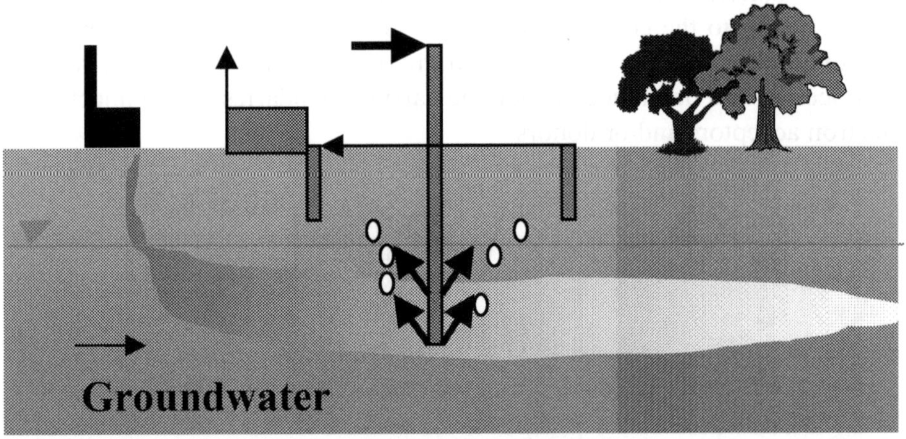

Figure 4. Injection of electron acceptors into a contaminant plume.

Table 4 presents the results of a batch test of a soil contaminated by 6.5g/kg dry matter of mineral oil, established to evaluate the feasibility of biosparging as a remediation technology (J. Gemoets, personal communication). Apart from a sterile control, air was added to two other microcosms. In one, additional nutrients were added. Only in the aerated situation, together with the addition of nitrogen and phosphorous, was any degradation observed (32%), and a peak O_2 uptake rate of 1.2g/kg/d was observed. This test clearly illustrated the benefit of nutrient supplementation, when this is lacking in the natural system, in addition to the need for sufficient electron acceptors.

Table 4. Batch tests measuring mineral oil biodegradation under biostimulation by air and nutrient supplementation.

Condition	Final mineral oil concentration (mg/kg)	% degradation	Max O_2 consumption rate (mg O_2/kg/d)	O_2 consumption rate at the end (mg O_2/kg/d)
Sterile control	6533	0	28	3
+ air	6233	5	91	51
+ air + N/P	4433	32	1200	56

Since oxygen availability is usually the key limiting factor for biodegradation, there is a significant need to develop alternative technologies for *in-situ* bioremediation of hydrocarbons under limited oxygen supply. Yerushalmi et al. (1999) and several other authors have described the biodegradation of mineral oil and BTEX in groundwater under micro-aerophilic conditions (oxygen concentrations below 2mg/l). During recent years, attention has been paid to the interface between the groundwater and surface water where water flows from poorly oxidized (groundwater) to highly oxidized (surface water) environments. In the transition zone, micro-aerophilic conditions create a biologically active zone where BTEX can be degraded. It is apparent that particularly the first reaction in the benzene biodegradation pathway is inhibited by anoxia.

4.2. ELECTRON DONOR REACTIONS

4.2.1. *Biodegradation of hydrocarbons*

Some compounds are highly oxidized (e.g., highly chlorinated hydrocarbons, nitro-aromatics and hexachlorohexane) and are readily reducible by the addition of an electron donor (Figure 5). Suitable electron donors can be: natural organic matter, HRC® (Hydrogen Release Compound), lactate, molasses, protamylasse and vegetable oils, and the hydrogen released by hydrogenotrophic micro-organisms during the fermentation of alcohols or volatile fatty acids. When chlorinated hydrocarbons are used as electron acceptors, the chlorine residue is sequentially replaced by hydrogen. These reactions occur most rapidly under sulphate-reducing or methanogenic conditions, although some reactions can also proceed under nitrate or iron-reducing conditions.

The hydrogen concentration determines the competition between this so-called halorespiration and methanogenesis. Acetotrophic bacteria such as *Desulfitobacterium* sp. are associated with incomplete dechlorination to cis-DCE while *Dehalococcoides* sp., which are hydrogenotrophic, are associated with complete dechlorination to ethene (Yang et al., 2005).

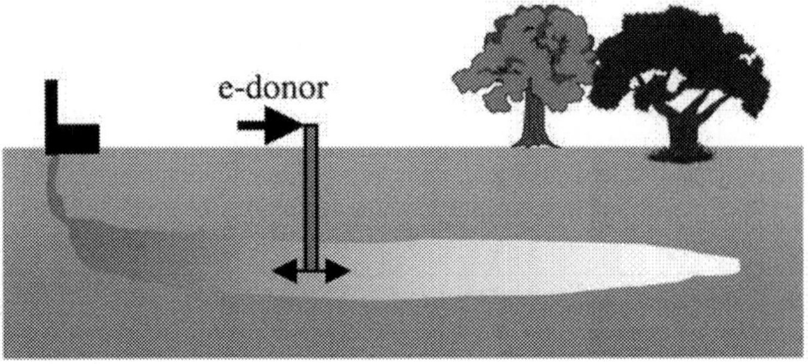

Figure 5. Injection of an electron donor into a groundwater contaminant plume.

Because of the high complexity of the processes involved, it is advisable to mount a feasibility study before any full-scale remediation process is attempted. Batch tests ('microcosms') are suitable, provided that they are conducted with representative field samples, and are run under strictly anaerobic conditions. Killed controls need to be included to distinguish biotic from abiotic dechlorination mechanisms. An example of a microcosm test is given in Figure 6. The test was conducted with material from a site polluted with PCE (ca. 100μM in the groundwater). Four conditions were tested: (1) a killed control; (2) with lactate added as an exogenous carbon source; (3) with Nutrolase® (protamylasse); (4) with Nutrolase and an added dechlorinating microbial inoculum. The results show that the site possesses good natural dechlorination potential, only requiring the addition of an external carbon source, for which lactate and Nutrolase were equally well-suited. Extra dehalogenating bacteria speeded the dechlorination, but were not necessary for the complete reduction to ethene.

4.2.2. *Bioprecipitation of heavy metals*

For *in situ* metal bioprecipitation (ISBP), sulphate-reducing bacteria are provided with electron donors such as molasses, lactate, HRC® (Koenigsberg et al., 2002), MRC® (Koenigsberg, 2002), ethanol and/or other carbon sources. The bacteria oxidize the electron donor and transfer

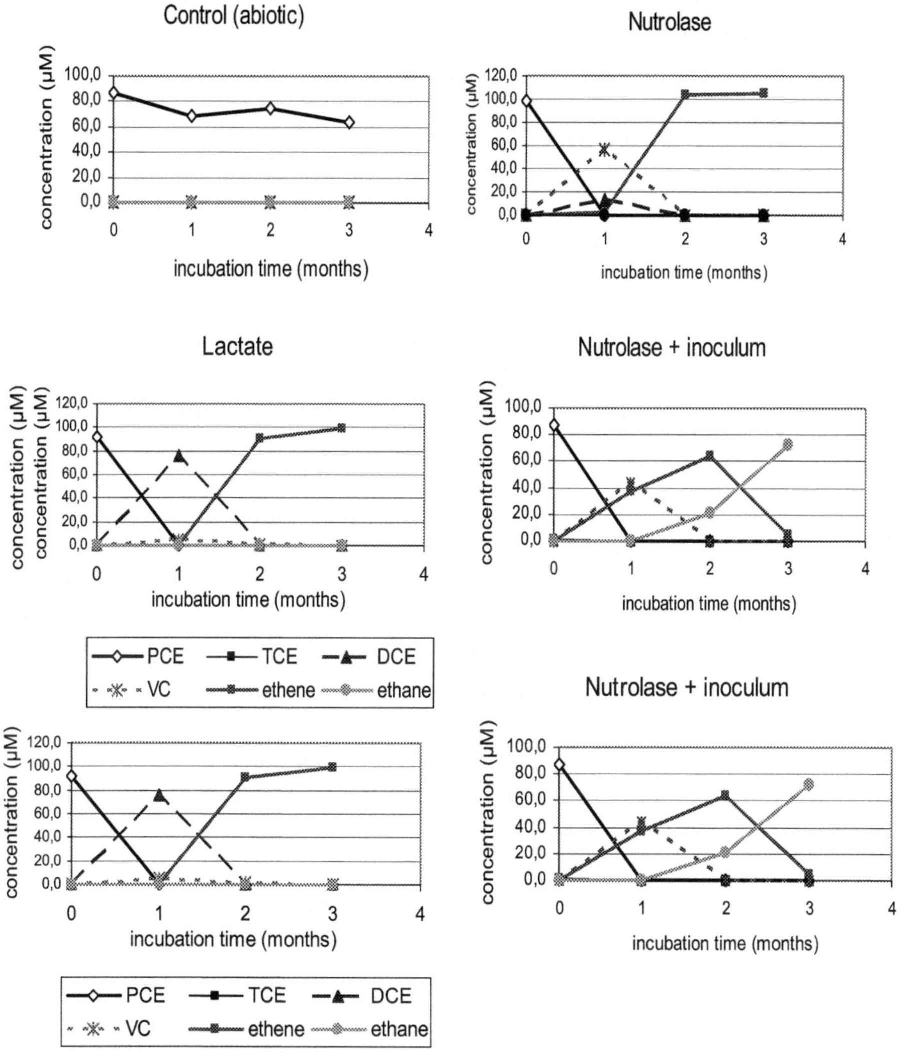

Figure 6. Example of results of a microcosm test investigating different conditions for dechlorination.

the released electrons to sulphate present in the water. Any metals present in the solution are precipitated in the form of sulphides (Janssen and Temminghoff, 2004). For the process to occur successfully, it is necessary that sulphate-reducing bacteria be present in the aquifer, sulphate be present at a sufficient concentration (generally >200mg/l), pH be in the range 5–8, a minimum content of nutrients (N and P) be provided, oxygen be absent, and a low redox potential prevails. The general principle is presented in the following reactions:

$$SO_4^{2-} + 8e^- + 8H^+ \rightarrow S^{2-} + 4\,H_2O$$

$$CH_3COOH + 2H_2O \rightarrow 2CO_2 + 8H^+ + 8\,e^-$$

$$CH_3COOH + SO_4^{2-} \rightarrow 2HCO_3^- + HS^- + H^+$$

$$H_2S + Me^{2+} \rightarrow \underline{MeS} + 2H^+$$

The process can be used to immobilize heavy metals such as Cd, Zn, Cu, Pb, Ni and Co. Cr(VI) can be reduced by some metal-reducing bacteria to the less toxic and less soluble form Cr(III). Arsenate [As(V)] can be reduced to the more mobile arsenite [As(III)] which precipitates as As_2S_3,, and is insoluble at low pH. Several laboratory-scale tests (batch and column) are currently available to study the feasibility of this process. However, only a few field tests have been performed to date. Two such tests have been conducted in Belgium, one at a non-ferrous industrial site, where the groundwater was contaminated with Cd, Zn, Ni and Co, and the other which was treated by injection of molasses in order to reduce chromium (VI) to chromium (III). A third demonstration in The Netherlands has been performed at a metal surface treatment site contaminated by Zn. The outcomes of a batch test of a groundwater heavily contaminated by Zn, Cd, Co and Ni are presented in Table 5. The initial sulphate concentration was 506mg/l. With the addition of acetate, a nearly

Table 5. Zn immobilization via *in-situ* bioprecipitation.

Condition	T0	T8	T20
	Zn concentration (µg/l)		
Aquifer + GW	101000	79200	49000
Aquifer + GW + $HgCl_2$	109000	94200	94200
Aquifer + GW + acetate	109000	82800	15
Aquifer + GW + 5 x acetate	103000	109000	90000
	Oxidation Reduction potential (ORP; mV)		
Aquifer + GW	-117	-60	-36
Aquifer + GW + HgC12	232	-50	-104
Aquifer + GW + acetate	-146	-70	-229
Aquifer + GW + 5 x acetate	-65	-78	-90

Zn concentrations are presented in µg Zn/l and redox is presented in mV.

complete immobilization of Zn in the groundwater was achieved (initial Zn level: 109mg/l; final level: 15μg/l). Under these conditions, the redox potential was -229mV, which is ideal for sulphate-reducing bacteria. At five times higher acetate concentrations, Zn immobilization and optimal ORP were not achieved, presumably due to acetate toxicity. More detailed results have been presented by Diels et al. (2005a, 2005b, 2005c).

5. Bioaugmentation

In certain cases, bacteria able to degrade the pollutants present are absent, or present in insufficient density in the natural soil environment. An instance of this situation has been noted involving pollution with MTBE (methyl *tert*-butyl ether), a gasoline additive. Many gasoline spills contain MTBE which migrates with the groundwater at a much higher rate than BTEX. Although MTBE-degrading bacteria exist (Fiorenza and Rifai, 2003), natural attenuation or biostimulation has frequently failed (Moreels et al. 2004). Many bacteria exist that can dehalogenate PCE or TCE to cis-DCE. Only very few species have been identified, however, that are able to reductively dechlorinate cis-DCE, via the carcinogen VC, to the harmless end product ethene. One of these species is *Dehalococcoides ethenogenes* (Maymo-Gatell et al., 1997). Lookman et al. (2005) have recently shown that rapid, full dehalogenation to ethene, after carbon source supplementation and without bio-augmentation, occurs only at a minority of sites. The critical factor is the presence of specific subtypes of *Dehalococcoides*, which express vinyl-chloride reductases (Krajmalnik-Brown et al., 2004).

In a test plot contaminated with MTBE to which oxygen was provided, Salanitro et al. (2000) have described the effect of the presence of an MC-100 MTBE-degrading consortium. In the natural attenuation plot, no degradation was detected, whereas in the oxygen provided plot (biostimulation) MTBE degradation was observed. However, the MTBE biodegradation in the consortium-amended plot showed a shorter lag time, and the extent of degradation was much higher than achieved in the oxygen-amended plot without the consortium.

Special attention must be paid to the injection of micro-organisms. The means of growing the biomass lead to the formation of slimes and extracellular polymers, resulting in biofilm growth on the injection well and finally to clogging of the injection system. Similarly, the injection of micro-organisms together with electron donor or acceptor can cause growth of biomass on the well and again lead to clogging. At the Dover Air Force Base in Delaware, clogging of injection wells was attributed to high levels of bacterial growth around the well opening (Grindstaff, 1998). Bacteria showed a tendency to accumulate near the opening of the well following the injection of substrate and nutrient-enriched groundwater. In order to solve these problems, both substrate and nutrients had to be pulse-injected in a continuous cycle. The injection schedule consisted of the injection over four days of substrate-enriched groundwater, followed by the circulation of unamended groundwater for half a day, the injection of nutrients and groundwater for another three days, and finally the circulation of unamended groundwater for half a day.

6. Summary and conclusion

Under proper conditions, monitored natural attenuation or enhanced attenuation (biostimulation or bioaugmentation) along with source removal, long-term monitoring, and land-use restrictions can be preferable to more expensive conventional technologies. At some sites, natural attenuation and/or bioremediation may not be possible because of regulatory constraints or unfavourable site conditions. The selection of the appropriate remedial technology at a particular site will be determined by time constraints for attaining the remediation objectives, the site's hydrogeology and geochemistry, the particular contaminants involved, prevailing regulatory constraints, and considerations of environmental exposure and cost.

References

Azadpour-Keeley, A., Russell, H. H., and Sewell, G. W., 1999, Microbial processes affecting monitored natural attenuation of contaminants in the subsurface, EPA/540/S-99/001.

Diels, L., Geets, J., Dejonghe, W., Van Roy, S., and Vanbroekhoven, K., 2005b, Heavy metal immobilization in groundwater by in situ bioprecipitation: comments and questions about carbon source use, efficiency and sustainability of the process. Consoil 2005, Proceedings of the 9th International FZK/TNO conference on Soil-Water systems, O. Uhlmann, G. J. Annokkée, F. Arendt (eds), 1622-1628.

Diels, L., Geets, J., Dejonghe, W., Van Roy, S., Vanbroekhoven, K., Szewczyk, A., and Malina, G., 2005c, Heavy metal immobilization in groundwater by in situ bioprecipitation: comments and questions about carbon source use, efficiency and sustainability of the process. Proceedings IMWA 2005 Mine closure, Departemento de Explotacion y Prospeccion de Minas (University of Oveido), Oveido (Spain). 5-7/09/2005, J. Loredo, F. Pendas (eds), ISBN 84-689-3415-1, p. 355-360.

Diels, L., Geets, J., Van Roy, S., Dejonghe, W., Gemoets, J., and Vanbroekhoven, K., 2005a, Bioremediation of heavy metal contaminated sites. In Soil Remediation series 6 (Proceedings of the European Summer School on Innovative Approaches to the Bioremediation of Contaminated Sites Soil Remediation series N°6, ed. Faba F., Canepa P., ISBN: 88-88214-33-X, 323-338.

Fiorenza, S., and Rifai, H. S., 2003, Review of MTBE biodegradation and bioremediation, *Biorem. J.* **7**:1-35.

Grindstaff, M., 1998, Bioremediation of chlorinated solvent contaminated groundwater, Prepared for US EPA Technology Innovation Office.

Hendrickx, B., Dejonghe, W., Brennerova, M., Cernik, M., Ledere, T., Buchelli, M., Bastiaens, L., Verstraete, W., Top, E., Diels, L., and Springael, D., 2005, Dynamics of an Oligotrophic bacterial aquifer community during contact with a groundwater plume contaminated with Benzene, Toluene, Ethylbebenzene and Xylenes: an in situ Mesocosm study, *Appl. Environ. Microbiol.* **71**:3815-3825.

Janssen, G. M. C. M., and Temminghoff, E. J. M., 2004, In situ metal precipitation in a zinc-contaminated, aerobic sandy aquifer by means of biological sulfate reduction, *Environ. Sci. Technol.* **38**:4002-4011.

Koenigsberg SS, Sandefur CA, Lapus KA, Pasrich GP (2002). Facilitated desorption and incomplete dechlorination observations from 350 applications of HRC. www.regenesis.com.

Krajmalnik-Brown, R., Hölscher, T., Thomson, I. N., Saunders, F. M., Ritalahti, K. M., and Löffler, F. E., 2004, Genetic identification of a putative vinyl chloride reductase in *Dehalococcoides* sp strain BAV1, Appl. Environ. Microbiol. **70**:6347-6351.

Lookman, R., Borremans, B., Gemoets, J., De Ceuster, T., Hamonts, K., Dejonghe, W., Bastiaens, L., Vanbroekhoven, K., and Diels, L., 2005, Chlorinated aliphatic hydrocarbons in aquifers: assessment of anaerobic degradation potential using microcosms and molecular biological techniques, In Consoil 2005, Proceedings of the 9th International FZK/TNO conference on Soil-Water systems, O. Uhlmann, G. J. Annokkée, F. Arendt (eds), 2139-2140.

Maymo-Gatell, X., Tandoi, V., Gossette, J., and Zinder, S. H., 1997, Characterization of an H2-utilizing enrichment culture that reductively dechlorinates tetrachloroethene to vinyl chloride and ethane in the absence of methanogenesis and acetogenesis, *Appl. Environ. Microbiol.* **61**:3928-3933.

Moreels, D., Bastiaens, L., Ollevier, F., Merckx, R., Diels, L., and Springael, D., 2004, Evaluation of the intrinsic MTBE biodegradation potential of European contaminated subsurface soils in microcosms simulating in situ bioremediation conditions, *FEMS Microbiol. Ecol.* 49:121-128.

Salanitro, J. P., Johnson, P. C., Spinnler, G. E., Maner, P. M., Wisnieuwski, H. L., and Bruce, C., 2000, Field scale demonstration of enhanced MTBE bioremediation through aquifer bioaugmentation and oxygenation, *Environ. Sci. Technol.* **34**:4152–4162.

Yang, Y., Pesaro, M., Sigler, W., and Zeyer, J., 2005, Identification of microorganisms involved in reductive dehalogenation of chlorinated ethenes in an anaerobic microbial community, *Water Res.* **39**:3954-3966.

Yerushalmi, L., Manuel, M. F., and Guiot, S. R., 1999, Biodegradation of gasoline and BTEX in a microaerophilic biobarrier, *Biodegradation* **10**:341-352.

PLANT TAXONOMY FOR PHYTOREMEDIATION

STANISŁAW W. GAWRONSKI AND HELENA GAWRONSKA*
Laboratory for Basic Research in Horticulture, Faculty of Horticulture and Landscape Architecture, Warsaw Agricultural University, Warsaw, Poland

*To whom correspondence should be addressed.

Abstract- Plants are among the most tolerant of organisms to pollution, which emphasises their utility for the emerging science of environmental biotechnology. Many botanical families, in particular the *Brassicaceae*, *Poaceae, Fabaceae, Asteraceae, Salicaceae, Chenopodiaceae,* and *Careophylaceae* include multiple species showing phytoremediation potential, and other families (*Cyperaceae, Amaranthaceae, Cannabaceae, Cannaceae, Typhaceae* and *Pontederiaceae)* contain promising individual species: Each species enjoys certain advantages, but suffers some limitations for application as phytoremediants. Careful selection of the appropriate family and genotype to match the particular pollutant and environment is crucial for successful phytoremediation.

Keywords: environmental pollution, heavy metals, hydrocarbons, air pollution, genetic diversity, phytoextraction, phytostabilization, phytomining

1. Introduction

Pollution of the soil, water and air, arising as a side effect of many modern human activities, needs to be reduced, and where possible, eliminated. Among the higher organisms, only plants have evolved effective strategies and mechanisms for survival in heavily polluted sites. They can thrive in environments contaminated to levels orders of magnitude greater that can be tolerated by other higher organisms. Plants cope with the presence of toxic metals in various ways, including their exclusion, their detoxification

N. Marmiroli et al. (eds.), Advanced Science and Technology for Biological Decontamination of Sites Affected by Chemical and Radiological Nuclear Agents, 79–88.

and their storage in specialised cells or cell compartments (vacuoles, cell walls). Two mechanisms for defence against heavy metal toxicity have been documented. One is based on the metallothioneins (Goldsbrough 2000) which are found in all living organisms; and the second on the synthesis of phytochelatins (Cobbett 2000, Rauser 1995), restricted to date to plants. The high capacity for the detoxification or degradation of pollutants shown by some plant species can be exploited for the emerging science of environmental biotechnology, termed phytoremediation. The concept of phytoremediation was inspired by the discovery of hyper-accumulators, most of which belong to the botanical families *Brassicaceae, Poaceae, Papilionaceae, Caryophyllaceae,* and *Asteraceae,* which provide most of the candidates for heavy metal phytoremediation. Two other families are important – the *Salicaceae* with the genera *Salix* and *Populus,* which are effective against a range of pollutants; and the *Betaceae* which contribute species effective against salt ions and small (few rings) polycyclic aromatic hydrocarbons (PAHs). Some *Asteraceae* species have been shown to be good phytoremediants of radionuclide pollution.

Some organic pollutants are readily taken up by plants and degraded. Examples of these are trichloroethylene (TCE) (Newman et al. 1997), pesticides and small (few aromatic rings) PAHs. Others, such as large PAHs (many rings), petroleum hydrocarbons (White et al. 2006), and polychlorinated biphenyls (PCBs) arc rather insoluble in water, and therefore are not taken up by plants roots in significant amounts. Plants, however, exert an indirect influence on the degradation of PAHs and PCBs by their interaction with particular rhizosphere micro-organisms. Thus, mulberry and sage orange (*Moraceae),* crabapple (*Rosaceae),* robinia locust *(Papilionaceae)* and birch (*Betulaceae)* can stimulate the degradation process by release into the rhizosphere of flavonoids (molecules containing six carbon rings), which encourages the growth of appropriate micro-organisms. It has been suggested that the similarity in chemical structure between flavonoids and PAHs / PCBs contributes to the stimulation of growth and activity of the degrading bacteria (Fletcher and Hegde 1995). Low molecular weight organic pollutants can also be present in gaseous form, and some of these are highly soluble in lipids, such as the waxes and cutin coating plant leaves, young shoots and tree bark, and thereby act as natural traps for this class of pollutant (Hermanson, Hites 1990, Staci, Hites 1994). Plant architecture and leaf hairs influence the dry deposition of PAHs, which are present in the atmosphere both in the gaseous phase and bound to particles. Small PAHs, with a molecular weight under 252, are predominantly present in the gaseous phase, while the larger ones are mainly particle-bound. Leaves which are very hairy appear to trap less of the smaller, and more of the larger PAHs, while glabrous leaves behave in the opposite way (Bakker et al. 1999).

A particular feature of phytoremediation is its suitability for the removal of common gaseous pollutants, such as CO_2, NO_2, CO and O_3, which is difficult to achieve by physical methods. Since CO_2 is a nutritional requirement for plant growth, elevated levels are favourable. Up to 10% of assimilated nitrogen is provided by the uptake of NO_2 in certain plant species (Morikawa at el. 1999). CO is metabolized mainly by micro-organisms, but to some extent also by plants (Orcutt and Nilsen 2000). Finally, O_3 is taken up and removed from the environment by certain plants which have a high capacity to scavenge hydroxyl radicals, and these are the species commonly recommended for cultivation in urban areas, where this pollutant is common.

2. Brassicaceae

Species of this family are among the best accumulators of heavy metals. Most of the documented hyper-accumulators belong to the genera *Alyssum* and *Thlaspi*. For phytoremediation, we have generally concentrated on domesticated species, and the favoured one in this family is Indian mustard (*Brassica juncea*). Some genotypes show a very high capacity to absorb heavy metals, and the crop can be cultivated twice a year. *Sinapis alba* – the white mustard commonly cultivated in Europe as a green manure – has a slightly lower phytoremediation capacity (Krysiak-Winska, Gawronski 2002). With the increasing focus on renewable fuels, the area sown to oilseed rape for the production of biodiesel is increasing significantly, and since the resulting oil is not destined for human or animal consumption, the cultivation of oilseed rape on polluted sites can serve the double purpose of industrial production and phytoremediation. Economic yields of rapeseed oil cannot, however, be obtained from very poor quality soils. Some disadvantages of using brassicas for phytoremediation are that measures need to be taken against insect damage, the dry leaf material is very fragile, and there is little interaction with symbiotic mycorrhizal fungi, the presence of which generally increases the level of tolerance to and uptake of heavy metals. The brassicas show particular promise as a means to recover valuable metals such as gold (*B. juncea* - Anderson et al.1998) and nickel (*Alyssum murale* – Chaney et al. 2005), referred to as “phytomining”.

3. Poaceae

The grass family is the one of the most important for the phytoremediation of heavy metal and organics such as the PAHs and petroleum hydrocarbons. An advantage of this family is that after cutting and drying, plant material is not fragile. Most species possess rather shallow root systems, but root density is high. The most productive biomass accumulator among the grain species is maize (*Zea mays*), which is tolerant to heavy metals and petroleum; among the temperate cereals, tall varieties of wheat (*Triticum aestivum*) and barley (*Hordeum vulgare*) have been recommended for phytoremediation. However, when a polluted site is saline, barley is the most tolerant of the cereals (Orcutt, Nilsen 2000). Among the domesticated fodder grasses, *Lolium* and *Festuca* are tolerant of PAHs and petroleum hydrocarbons, as demonstrated by their frequent presence on polluted road-sides. *Lolium* in Europe is represented by *L. perenne* and *L. multiflorum*, and *Festuca* by *F. rubra, F. arundinacea* and *F. ovina.* varieties from both genera have been recommended for mixed sowing along with legumes in areas polluted by petroleum hydrocarbons (White et al. 2006). Pasture grasses produce less biomass than grain crops, but they can be cut several times during the season, allowing absorbed pollutants to be removed before they are eluted from the plants by rain.

The genus *Agrostis* features two highly resistant species *A. tenuis* and *A. alba,* which have a particularly high level of resistance to heavy metals. However, because this is achieved by restricting pollutants to the roots, they are suitable for phytostabilization, rather than for phytoremediation. A frequent coloniser of polluted sites in the temperate zone is *Agropyron repens,* which is very tolerant to salinity and heavy metals, accumulating them in high amounts in underground rhizomes which can be mechanically pulled from the soil. Two further grasses with a high phytoremediation capacity are cultivated in Europe as ornamental species – these are *Deschampsia cespitosa* and *Vetiveria zizanioides.*

For the phytoremediation of water and sediments, the prime temperate species is *Phragmites australis,* which is very tolerant to a wide range of pollutants, including high salinity. Other species such as *Phalaris arundinacea* and *Miscanthus sp* are adapted to warmer climates. *Miscanthus giganteus* accumulates a large amount of biomass, and within two years is sufficiently frost tolerant to survive the Central European winter. Grasses are associated with a wide range of micorrhizal fungi, especially *Glomus* spp., with *G. intraradices* particularly suited to heavy metal polluted sites.

4. Fabaceae

Many Fabaceae species are good phytoremediants of heavy metal pollution, while their leaf and bark waxes trap organic pollutants, and are effective for the stimulation of growth of rhizophere micro-organisms capable of PAH and PCB degradation. The advantages of this group is their self-sufficiency in terms of nitrogen supply, and their favourable level of tolerance to drought. Some species are adapted to very poor soils. The absorption of heavy metals by legumes has been well documented (Piechalak et al. 2002). Some *Trifolium* spp., *Medicago sativa* and *Melilotus officinalis* are effective, in combination with grasses for the phytoremediation of petroleum hydrocarbons. In urban areas, ornamental trees and shrubs from this family such as *Robinia pseudoacacia, Caragana arborescens* and *Amorpha fruticosa* are commonly cultivated. All three species efficiently absorb heavy metals and trap volatile organic pollutants. *R. pseudoaccacia* exudes significant quantities of flavonoids to the soil. These trees and shrubs recover well from pruning, which allows for the ready removal of polluted plant material. Several hyper-accumulators of selenium have been described in the genus *Astragalus* (Brooks 1998).

5. Asteraceae

Species from this family have been used for the bio-removal of heavy metals and radionuclides, such as Sr, Cs and U. High biomass varieties of annual *Helianthus annus* and perennial *H. tuberosus* are particularly suitable for the phytoremediation of polluted industrial sites. The former can accumulate as much as 100t/ha biomass, is tolerant to drought and highly competitive against weeds. As the latter is a perennial crop, tilling the polluted soil can be avoided. However, biomass productivity decreases over time, and eventually replanting does become necessary. Its tubers are sought after by wild pigs, so cultivation is not recommended near forest margins. The Asteraceae are rich in ornamental species and some accumulate high amounts of biomass and may thus be candidates for phytoremediation in urban sites. Species from the genera *Solidago, Tanacetum* and *Rudbeckia* are particularly suitable in this context. Typically tall, large biomass producers are preferred, even if these are not always aesthetically the most pleasing. Some species, in particular *Artemisia vulgaris,* have shown good tolerance to salinity.

6. Salicaceae

A number of the woody fast growing species in this family produce large biomass and absorb a wide range of pollutants. These properties fit well with the requirements for phytoremediation. Fast-growing trees are typically short-lived, but this is not an issue in phytoremediation. The genus *Salix* contains several useful species. *S. viminalis* is well suited for heavy metal (Cd, Pb and Zn) phytoextraction and biomass production for renewable energy. Significant genetic variation with respect to phytoremediation ability has been observed in this species, and particular varieties of *S. viminalis* are available for heavy metal uptake. The species is very easily vegetatively propagated, and responds well to irrigation. Plantations can survive more than ten years, but during first two to three years, weed control is required.

The shrub or small tree *S. caprea* has good phytoremediation ability, but is less commonly used. This is largely because it is seed propagated, and is associated with genetic instability. In recent years, *S. burjatica* has also been recommended as a candidate for phytoremediation in locations where tolerance to low temperature is important (Pulford et al. 2002). Willows intercross readily, and many commercially offered varieties are hybrids (Pulford et al. 2002). One of the oldest such crosses is *S. viminalis* x *S. caprea,* known as *S. smithiana*. This hybrid grows very fast, accumulates high biomass, absorbs heavy metals efficiently, and is easily propagated by cuttings. *S. viminalis, S. daphnoides* and *S. pupurea* are also cultivated as a ornamental species, and in urban areas polluted by heavy traffic provide a level of phytoremediation of air pollutants. PAHs, PCBs and dioxins are accumulated in leaf and stem waxes, so removal of leaf litter in the autumn and pruning in the spring together act to remove some these carcinogenic pollutants from the urban environment.

Populus is used for the phytoremediation of heavy metals, and certain organic pollutants. By the age of five years, roots can extend to 20m, transpiring up to 200 litter of water per plant per day. In the temperate zone, poplars are one of the fastest growing of plant species and accumulate a large biomass. They can easily be propagated by cuttings. The most successful genotypes are not species, but rather are interspecies hybrids, which can maintained by vegetative propagation. European x American hybrids are common, such as *P. nigra var italica* x *P. deltoides* (also known as *P. euroamericana*), *P. trichocarpa* x *P. deltoides* and *P. trichocarpa* x *P. maximowiczi.* The rapid growth habit of this species requires a plentiful supply of water, but also means that competition from weeds is unimportant. Poplar is particularly suited for the remediation of trichloroethylene (TCA), because a deep root system is necessary to recover this compound which, being heavier than water, migrates below the underground water. The poplar can also absorb and either degrade or directly transpire TCE (Newman et al. 1997). Both *Salix* and *Populus* spp. are readily colonized by mycorrhizal fungi, which increase the hosts phytoremediation efficiency (Sell et al. 2005).

7. Chenopodiaceae

Some chenopods are highly tolerant to salinity, and are thus an appropriate choice for the rescue of salt polluted sites. They have also been recognized as efficient absorbers of radionuclides, and are capable of degrading small PAHs (Harms et al. 2003). Various beets are commonly cultivated as high biomass crops, but have not been used for phytoremediation because of problems surrounding the treatment and disposal of the harvested biomass. Two other cultivated species accumulating high biomass are *Atriplex hortensis* and *Kochia scoparia,* both ornamental species which are easy propagated from seed. A number of varieties have been commercially selected. Few chenopod species are colonized by mycorrhizal fungi.

8. Careophylaceae

Species from this family are often found on salt or heavy metal polluted sites. Since they accumulate little biomass they are not used for phytoremediation, but their presence is taken as an indicator of pollution.

9. Cyperaceae

This family possesses several species very tolerant to pollution. Many species are adapted to very wet sites, typically on the edge of bodies of water, and so are suited for phytoremediation in such environments. In the temperate and cooler zones of Europe, *Carex hirta* can be recommended; *Cyperus alternifolius* is an alternative for more southerly parts of the continent, and similarly, *C. papyrus* for the Mediterranean and subtropical zones. All these species can be vegetatively propagated by division of the mother plant.

10. Amaranthaceae

Amaranth (*Amaranthus* spp.) is cultivated as an ornamental plant, and also of recent years as an edible non-cereal seed. Many interspecies hybrids have high tolerance to salinity and the capacity to absorb heavy metals.

Fast-growing and high-biomass-accumulating varieties can grow up to 2m in height, and these are suitable for the phytoremediation of salt, heavy metal or both.

11. Cannabaceae

Cannabis sativa shows much promise as a phytoremediant. Individual plants can grow up to 4m high in a season, accumulating a large biomass. Although perennial, the herbaceous biomass is can be fully harvested in the autumn, and the same plant can be cultivated over three to four years. To achieve high biomass yield, fertilizer application and irrigation is normally required. The species is a good phytoremediant for heavy metals and radionuclides, and significant differences in phytoremediation capability between varieties have been noted. Because of its close relationship with *C. indica,* its cultivation in countries with strong anti-drug laws remains problematical.

12. Cannaceae

Canna x *generalis* is one of the most common ornamental species in European cities. It has good potential for phytoremediation with respect to uptake and accumulation of heavy metals. Plants grow to 2m, and the leaves and stems are covered with a thick layer of wax, which traps insoluble gaseous organic pollutants such as PAHs, PCBs and dioxins. *Canna* varieties differ in their capacity for pollutant uptake, but since the species is generally vegetatively propagated, it is easy to maintain this trait. *Canna* x *generalis* was bred from crosses between species adapted to marshy environments, and thus favours moist soils. In warmer regions, *C. flaccida* is recommended for the phytoremediation of water.

13. Typhaceae

The most common European aquatic plants are the two water lily species *Typha angustifolia* and *T. latifolia.* The former produces more biomass, although their co-cultivation is encouraged to maintain species diversity. Both *Typha* and *Phragmites* favour still water, and both suppress algae when the leaf cover exceeds 25% of water surface. *Typha* phytoremediates heavy metals and agricultural pollutants such as nitrogenous fertiliser and pesticides. Plants translocate the pollutants to the upper part of the plants more efficiently than does *Pragmites,* but the latter species are more tolerant to salinity.

14. Pontederiaceae

The water hyacinth *Eichhornia crassipes* originates from Brazil, and although it represents one of the most troublesome and invasive of water weeds, it is also one of the best phytoremediants of polluted water. The biomass becomes saturated with pollutants, which can then be removed from the water by harvest of the biomass and its composting, a process which achieves a substantial reduction in the mass requiring further processing (e.g., incineration under controlled conditions). The species is highly susceptible to frost, but in frost-free environments, its introduction is risky as it can readily become uncontrollable.

The list we have presented is far from complete, and is developing as our knowledge of the plant kingdom increases. A recent exciting discovery has been Ma et al.'s (2001) description of the fern *Pteris vittata,* which hyper-accumulates arsenic, and thus may be developed as a phytoremediation technology for this particularly dangerous pollutant.

References

Anderson C. W. N., Brooks R. R., Stewart R. B., and Simcock R., 1998, Harvesting a crop of gold in plants, *Nature* **395**:553-554.

Bakker M. I., Vorehout M., Sum D. T. H. M., and Kolloffel C., 1999, Dry deposition of atmospheric polycyclic aromatic hydrocarbons in three *Plantago* species. *Environ. Toxicol. Chem.* **10**: 289-2294.

Brooks R. R., 1998, *Plants that hyperaccumulate heavy metals*, CAB International, University Press, Cambridge.

Chaney R. L., Angle J. S., Wang A. S., McIntosh M.S., Broadhurst L., and Reeves R. D., 2005, Phytoextraction of soil Cd, Ni and Zn using hyperaccumulator plants to alleviate risks of metal contaminated soils requiring remediation. International Workshop Current developments in remediation of contaminated lands p. 39, 27-29 October 2005. Pulawy, Poland.

Cobbett C. S., 2000, Phytochelatins and their role in heavy metal detoxification, *Plant Physiol.* **123**:825-832.

Fletcher J. S., and Hegde R. S., 1995, Release of phenols by perennial plant roots and their potential importance in bioremediation. *Chemosphere* **31**:3009-3016.

Goldsbrough P., 2000, Metal tolerance in plants: The role of phytochelatins and metallothioneins, in: *Phytoremediation of contaminated soil and water*, N. Terry, G. Banuelos, eds, Lewis Publishers, Boca Raton.

Harms H., Bokern M., Kolb M., and Bock C., 2003, Transformation of organic contaminants by different plant system, in: *Phytoremediation; Transformation and control of contaminants*, S. C. McCutcheon, J. L. Schnoor, eds., John Wiley & Sons, Inc., Hoboken, New Jersey.

Hermanson M. H., and Hites R. A., 1990, Polychlorinated biphenyls in tree bark, *Environ. Sci. Technol.* **24**:666-671.

Ma Q. L., Komar K. M., Tu C., Zhang W., Cai Y., and Kennelley E. D., 2001, A fern that hyperaccumulates arsenic, *Nature* **409**:579.

Newman A. L., Strand S. E., Choe N., Duffy J., Ekuan G., Ruszaj M., Shurtleff B. B., Wilmoth J., Heilman P., and Gordon M. P., 1997, Uptake and biotransformation of trichloroethylene by hybrid poplar, *Environ. Sci. Technol.* **31**:1062-1067.

Morikawa H., Higaki A., Nohno M., Takahashi M., Kamada M., Nakata M., Toyohara G., Okamura Y., Matsui K., Kitani S., Fujita K., Irifune K., and Goshima N., 1999, More than a 600-fold variation in nitrogen dioxide assimilation among 217 plant taxa, *Plant Cell Environ.* **21**:180-190.

Orcutt D. M., and Nilsen E.T., 2000, *The physiology of plants under stress*, John Wiley & Sons Inc., New York.

Piechalak A., Tomaszewska B., Baralkiewicz D., and Malecka A., 2002, Accumulation and detoxification of lead ions in legumes. *Phytochemistry* **60**:153-162.

Pulford I. D., Riddel-Black D., and Stewart C., 2002, Heavy metal uptake by willow clones from sewage sludge-treated soil: The potential for phytoremediation. *Int. J. Phytorem.* **4**:59-72.

Rauser W. E., 1995, Phytochelatins and related peptides, *Plant Physiol.* **109**:1141-1149.

Sell J., Kayser A., Schulin R., and Brunner I., 2005, Contribution of ectomicorrhizal fungi to cadium uptake of poplar and willows from a havey polluted soil, *Plant Soil* **277**:245-253.

Staci L. S., and Hites R. A., 1994, Importance of vegetation in removing polycyclic aromatic hydrocarbons from the atmosphere, *Nature* **370**:49-51.

White P. M., Wolf D. C., Thoma G. J., and Reynolds C. M., 2006, Phytoremediation of alkylated polycyclic aromatic hydrocarbons in a crude oil-contaminated soil, *Water Air Soil Pollut.* **160**:207-220.

Winska-Krysiak M., and Gawronski S. W., 2002, Fizjologiczne aspekty tolerancji i hiperakumulacji ołowiu w wybranych genotypach *Brassica.* [Physiological aspects of lead tolerance and hyperaccumulationin *Brassica* species]. *Zeszyty Problemowe Postepów Nauk Rolniczych* **481**:605-613.

Wolverton B. C., 1997, *How to grow fresh air: 50 house plants that purify your home or office*, Penguin. 144 p. New York. USA.

GENETIC VARIABILITY AND GENETIC ENGINEERING IN PHYTOREMEDIATION

NELSON MARMIROLI*
Department of Environmental Sciences, University of Parma, Viale G.P. Usberti 11A, 43100 Parma, Italy

*To whom correspondence should be addressed.

Abstract- Plants adapted to growth in contaminated environments can be exploited for the purpose of phytoremediation. In order to evaluate their potential, genetic approaches are required to uncover the allelic variation present at both Mendelian and quantitative genes. Natural populations can provide a wealth of information regarding the mechanisms of tolerance and the accumulation of heavy metals, and recent evidence has established the genetic independence of these two aspects of the plant response to metal toxicity. Model plant systems provide an ideal tool to develop genomics and proteomics approaches in this area. The identification of significant genes, and an understanding of their properties are important for the development of genetic engineering strategies aimed at improving plants for their phytoremediation ability. Transgenic plants for metal remediation are rather inefficient and display contradictory behaviour, probably because metal accumulation is a multigenic trait. On the other hand, the insertion of single genes can confer novel degradation properties targeted at specific organic contaminants, and several such examples are described.

Keywords: genetic engineering, genetic diversity, heavy metals, metal accumulation, metal tolerance, model plants, organic contaminants, quantitative trait loci

1. Genetic variability in phytoremediation

Plants growing in contaminated environments can be studied and their behaviour analysed in order to obtain information useful for their exploitation in phytoremediation. Much variability can be observed with respect to both phenotype and adaptation, but not all of this variability has a genetic basis. Plants transplanted from their native environment may

N. Marmiroli et al. (eds.), Advanced Science and Technology for Biological Decontamination of Sites Affected by Chemical and Radiological Nuclear Agents, 89–108.

change their behaviour and phenotype, even though the genotype is fixed, a phenomenon referred to as phenotypic plasticity (Via et al., 1995). Formally, phenotypic plasticity describes the way in which a plant's form is determined by the environmental conditions in which it finds itself. Exploitation of phenotypic plasticity is difficult because the factors that govern it are unknown and cannot be controlled. Only those properties which depend on the genetic make-up of the individual plant can be reliably exploited in phytoremediation, because the genes are present in all environments. However, their expression can be modified by the environment, so for exploitation, it is necessary to define the environmental conditions which do not exert too high a pressure, otherwise genetically-based features will fail to be expressed in a reliable way.

1.1. APPROACHES FOR THE STUDY OF GENETIC VARIABILITY

Two types of genetic systems prevail. The first is Mendelian, where a trait is determined by allelic variation at a single (or a small number) gene, so that its inheritance follows simple Mendelian rules; and the second is polygenic, where the trait (generally quantitative in nature) is determined by a number of genes and is generally subject to modification by the environmental conditions, so that phenotypic variability reflects contributions from both genetic and environmental factors. Mendelian systems allow the definition of discrete phenotypic classes, whereas quantitative systems are characterised by a continuous distribution of phenotype. The identification and isolation of a Mendelian gene is laborious but feasible, especially with the aid of modern genomics. However, the identification and isolation of genes responsible for quantitative effects, commonly referred to as quantitative trait loci (QTL), is much more challenging. A common starting point for their isolation lies in defining, with some degree of certainty, their genetic map position to a specific genomic region. In some cases, the coincidence of the map location of a QTL with that of a candidate Mendelian gene provides a short-cut to the genetic dissection of the trait (Paterson et al. 1988).

In phytoremediation, genetic approaches have been used to unravel the biological mechanisms underlying the response of plants to inorganic and organic contaminants, as a complement to physiological, biochemical and molecular biological approaches. The most intensively studied topic in this context is the accumulation of heavy metals. The long term aim of many of these efforts is to define the genetic determinants of accumulation, and to use these as a guide for either plant selection or as the basis for the design of phytoextraction technologies. Some mechanisms for tolerance and accumulation of metals have been described in the public domain (Yang et al., 2005; see Figure 3 in Pilon-Smits, 2005 for a summary), and most involve, at various levels, active transport across membranes, chelation, and sequestration in specific cell compartments.

1.2. MODEL PLANTS AND SYSTEMS IN GENETIC STUDIES OF METAL ACCUMULATION

Model plant systems have been favoured as the most effective means of understanding the genetic basis of metal accumulation. Most studies focus on the two metal accumulating species *Thlaspi caerulescens* J. et C. Presl. and *Arabidopsis halleri* (L.) O'Kane and Al Shehbaz. Both species belong to the Brassicaceae family, which has allowed direct comparisons to be made with *A. thaliana* (L.) Heynh, taking advantage of the good level of conservation of genetic structure and the ready exploitation of its genomic resources (www.arabidopsis.org). In both species, it is possible not only to perform crosses between individuals with contrasting phenotypes for tolerance or for accumulation, but also to produce crosses between species.

1.3. INTERSPECIFIC VARIABILITY FOR METAL ACCUMULATION AND TOLERANCE

Interspecific variation is best analysed by comparing plants from related species grown in a similar environment. Further studies can then be performed by inter-crossing individuals of sexually compatible species, obtaining an F1 hybrid progeny, and self-fertilising this hybrid to generate a population in which parental characteristics segregate. Alternatively, the F1 hybrid may be back-crossed to one (or both) of its parents, to obtain the expression of recessive traits. The segregation of phenotypes in the progeny predicts the genetic segregation in the gametes of the F1 hybrid. The individuals used for comparative studies and/or interspecific crosses are generally non-tolerant and/or non-accumulators (*T. arvense* L. for *T. caerulescens*; *A. lyrata* ssp. *petraea* (L.) O'Kane and Al Shebhaz for *A. halleri*).

1.3.1. *Arabidopsis halleri*

A. halleri is found on sites with high levels of Zn, Cd and Pb, but while plants constitutively accumulate Zn and Cd, they do not accumulate Pb (Bert et al. 2002). Individuals from non-metallicolous (i.e., environments where there is no metal stress) populations have a higher capacity to hyper-accumulate than do plants adapted to contaminated sites. Among back-cross progenies from the hybrid *A. halleri* x *A. lyrata,* segregation was observed for tolerance and accumulation (Bert et al., 2003). Tolerance to Cd acts as a semi-dominant trait, and the distribution in the progeny suggests that more than one gene is involved. Cd tolerance and accumulation segregate as independent traits, since different classes of tolerant individuals accumulate the same amounts of Cd. Both tolerance and accumulation of Cd and Zn co-segregate with one another, indicating that these traits are regulated by a common genetic system, either by the same genes or by distinct genes within a linkage block. Modifiers, acting on the level of tolerance, could also be involved.

The potential of genetic analysis has been shown by studies of a specific gene, *MTP1* (metal tolerance protein 1) (Dräger et al., 2004). This gene encodes a cation diffusion facilitator protein and is highly expressed in leaves and roots of *A. halleri*. In roots it is up-regulated by exposure to high Zn concentration. The genome of *A. halleri* contains three unlinked copies of *MTP1*, two of which co-segregate with zinc tolerance in the back-cross progeny. Individuals with both genes from the *halleri* parent have higher tolerance, while individuals carrying only one gene have intermediate tolerance. The non-tolerant species *thaliana* and *lyrata* possess only one copy. Thus *MTP1*seems to be a good candidate for the dissection of QTL for tolerance in this and perhaps also other species.

1.3.2. *Thlaspi caerulescens*

In a similar approach, the Zn accumulation in *T. caerulescens* has been compared with that of *T. arvense* (Assunçao et al., 2001). Genes encoding zinc transporters have been isolated from *caerulescens* and their expression levels studied in various conditions. *ZNT1* and *ZNT2* in roots, and *ZTP1* in leaves, are more highly expressed than in *arvense*; in the latter species, *ZNT1* and *ZNT2* are expressed only in conditions of Zn deficiency. This suggests that hyper-accumulation depends on an alteration in the transcriptional control of genes involved in metal transport across membranes. Only at very high metal concentrations are these gene down-regulated in *caerulescens* plants. These studies provide indications for specific genes which are differentially regulated in similar species and probably involved in tolerance and accumulation. Further genetic and physiological investigations are now needed to confirm their roles.

1.4. INTRASPECIFIC VARIABILITY IN METAL TOLERANCE AND ACCUMULATION

The same comparative tools used for assessing variability between species are equally applicable for the study of intraspecific variability.

1.4.1. *Thlaspi caerulescens*

T. caerulescens is generally tolerant of heavy metals, and hyper-accumulates Zn, but genetic variation for these characters has been documented within the species. Subspecies *calaminare* is commonly found in nature on Zn-rich soils, whereas subspecies *caerulescens* is more typical of normal soils (Meerts and van Isacker, 1997). Pot experiments have shown that all *calaminare* types have greater tolerance to heavy metals, whereas for some *caerulescens* ones, growth was inhibited by the presence of metals. Thus it is likely that above a background level of constitutive tolerance, natural selection has led to the evolution of specific tolerance mechanism(s). With regards to accumulation, *caerulescens* accumulates Zn to higher levels than does *calaminare*. The evidence is that accumulation is not correlated with tolerance, although the correlation does apply in some other species (Yang et al., 2005). In general, *caerulescens* seems to be the more variable for both Zn tolerance and accumulation.

In crosses between genotypes selected from soils with normal levels of metal, progeny accumulate more Zn than do those derived from parents sourced from soils contaminated with high levels. This has led Frérot et al. (2003) to propose that the major control of this character is exerted through a single gene, with a dominant allele restricting Zn hyper-accumulation. As there was no variation for Cd accumulation among the same individuals, the mechanisms involved are clearly independent. Similarly, by comparing extreme accumulation segregants, Assunçao et al. (2003) demonstrated that tolerant segregants accumulate less Zn than do chlorotic (sensitive) ones. This negative correlation may be due to linkage of genes with contrasting effects, or may reflect the presence of pleiotropic effects of a single gene. However, the continuous distribution of accumulation phenotypes among the progeny indicates the segregation of multiple genes. Using crosses between high and low accumulating ecotypes, Zha et al. (2004) showed that the accumulation of Cd and Zn is controlled by multiple genes, and that for Cd, accumulation and tolerance are independent of one another. Assunçao et al. (2001) compared *T. caerulescens* individuals selected from different populations for the expression of genes critically involved in Zn transport, and showed that *ZTP1* expression is higher in plants from calamine soils (rich in Zn) than in plants derived from serpentine or normal soils. Thus

ZIP I represents a potential candidate gene for the phenotypic segregation seen in many of the above crosses.

1.4.2. *Arabidopsis halleri*

Genetic variability for Zn accumulation in *A. halleri* has been described in 17 European populations (Macnair, 2002). At high Zn concentrations, all individuals hyper-accumulate, independently of provenance, in contrast to the experience with *T. caerulescens*. At low Zn concentrations, considerable variation exists, once again uncorrelated with selection site. From genetic analyses of intraspecific crosses, the heritability of Zn accumulation was estimated to lie in the range 0.25-0.5, and there was a significant positive correlation between Zn concentration at the maternal parent collection site and the accumulation capacity of the progeny.

1.5. QUANTITATIVE GENETICS FOR METAL ACCUMULATION AND TOLERANCE

Several of the experiments outlined above suggested the involvement of more than one gene in the determination of hyper-accumulation and tolerance, although in some cases, Mendelian genes were identified in crosses between divergent genotypes. Where quantitative inheritance applies, QTL mapping is used to identify the genetic basis of thc trait. Several research groups are constructing such maps in both model species and hyper-accumulators. Recent progress has been described for *T. caerulescens* (Denieau et al., 2004) and *A. halleri* (Willems and Samitou-Laprade, 2004). The former mapping population involves a cross between La Calamine and Ganges, accessions which contrast for Cd and Zn accumulation. In *A. halleri* three major Zn tolerance QTL have been mapped, each contributing equally, and overall explaining 33% of the total variance for this trait. Whether any of these QTL co-locate on the genetic map with the gene *MTP1*, which has been associated with Zn tolerance (Dräger et al., 2004) remains to be seen.

1.6. NEW TRENDS AND APPROACHES

Clearly, new genes involved in tolerance and hyperaccumulation remain to be discovered. Current approaches exploit genomics, proteomics and metabolomics to compare contrasting accessions and/or species. These approaches are greatly facilitated by the synteny and homology between the test and model species. When *A. thaliana m*icroarrays were used to compare gene expression patterns of *A. halleri* grown under different environmental conditions, a number of transcripts induced by exposure of

A. thaliana to Zn were found to be constitutively expressed in *A. halleri*, including putative Zn transporters, cation diffusion facilitators, nicotianamine synthase, cysteine synthase, transporters of the HMA family and *MTP1* (Becher et al., 2004; Weber et al., 2004). Real-time RT PCR analysis confirmed the higher expression levels for some of these genes, and mass spectrometry analyses of the proteome showed that the relevant proteins were present in higher amounts. These results suggest that elevated tolerance may require the constitutive expression of metal homeostasis genes in shoots. Similarly, array analyses of segregating progenies have identified the differential expression of about 140 genes in response to the presence of elevated levels of Cd (Courbot et al., 2004). Some of these are constitutively expressed in tolerant genotypes, whereas in sensitive genotypes they are induced by Cd. Thus it appears that a high level of constitutive expression of transcripts from metal detoxification genes may confer the ability to quickly detoxify metal ions under a fluctuating metal supply. *A. thaliana* arrays have also been exploited for profiling gene expression in *T. caerulescens*. These confirm the constitutive expression of genes in the hyper-accumulator which are induced or repressed by Zn in the non-hyper-accumulator (van de Mortel et al., 2004). Interestingly, however, 20% of *Thlaspi* spp. genes have no orthologues in the *A. thaliana* genome. Future breakthroughs in the elucidation of the genetic basis of metal tolerance and hyper-accumulation will no doubt emerge from genomics and proteomics.

2. Genetic engineering in phytoremediation

Genetic engineering is a mechanistic approach, in which gene function is studied by its expression in a different context. The rationale is that a gene's properties will be conserved even when transferred into a new organism and/or to a new genetic background. The appearance of a new property in a transgenic individual therefore allows the gene function to be identified. Successful transgenics may of course be of direct application in practical phytoremediation.

In classical genetic approaches, variability can be studied with more or less difficulty. Results can be problematic to interpret, but the success of the experiment is certain. In contrast, the success of a genetic engineering approach depends on the efficiency with which the transgene is expressed in its new genetic context. Particular attention must be paid to the design of the expression cassette, including codons, promoter sequences, terminators and introns. Genetic engineering approaches have addressed the metabolism and remediation of organic contaminants and inorganic contaminants in various ways (Cherian and Oliveira, 2005). In the case of organic contaminants, most plants have been engineered by the insertion of enzymes from different organisms to confer novel catabolic properties. These enzymes have to act within a complex network of existing enzymes co-adapted to work together. In the case of inorganic contaminants, most approaches have addressed transport proteins, chelators and tolerance-related compounds, with different levels of involvement in tolerance and accumulation. Genetic engineering approaches complement the "omics" technologies: the latter produce a catalogue of genes, proteins, and metabolites, whereas the former explores the function of a specific gene in the global context. Two basic genetic engineering strategies are relevant to improve phytoremediation: transformation with plant genes which contribute to the enhancement of existing properties; and the transfer of novel properties via genes from other species, including animals, microorganisms and other plant species. Table 1 collates some examples of transgenic organisms, which will be discussed below.

Table 1. Summary of recent research on transgenic plants for phytoremediation.

Candidate gene	Metal tolerance	Metal accumulation	Organics tolerance	Organics metabolism
NtCBP4	Increased for Ni, decreased for Pb	Higher for Pb, lower for Ni		
HMA4	Increased for Zn, Cd	Increased for Zn, Cd		
Metallothioneins	Increased for Cd			
ATP phosphoribosyl transferase	Increased for Ni			
ATP sulfurylase	Increased for Se	Increased for Se		
Cysteine synthase	Increased for Cd, Se, Ni	Increased for Cd		
Glutathione synthase	Increased for Cd	Increased for Cd		
γ–glutamylcysteine synthetase	Increased for Cd	Increased for Cd, Zn	Increased fror herbicides	Increased fror herbicides
YCF1	Increased for Cd, Pb	Increased for Cd		
Selenocysteine	Increased for	Increased for		

Table 1. Continued.

methyltransferase	Se	Se		
Selenocysteine lyase	Increased for Se (decreased if chloroplast expression)	Increased for Se		
Arsenate reductase (and γ–glutamylcysteine synthetase)	Increased for arsenate, Cd	Increased for arsenic, Cd		
ZntA	Increased for Pb, Cd	Decreased for Pb, Cd		
AtNramp3	Decreased for Cd	Increased for Fe, unchanged for Cd		
Phytochelatin synthase	Decreased for Cd			
Pentaerythritol tetranitrate reductase			Increased for GTN, TNT	Increased for GTN, TNT
Nitroreductase			Increased for TNT	Increased for TNT
Cytochromes P450			Increased for TCE, herbicides	Increased for TCE, herbicides
Chlorocatechol dioxygenase			Increased for 3-chlorocatechol	Increased for 3-chlorocatechol
Laccase			Increased for phenolics	Increased for phenolics
MnP peroxidase			Increased for PCP	Increased for PCP

2.1. INORGANIC CONTAMINANTS

2.1.1. *Transport systems*

As shown above, genetic and genomic evidence suggest that membrane proteins for metal transport are major determinants of accumulation. NtCBP4 is a putative cation channel of the plasma membrane of *Nicotiana tabacum* capable of binding calmodulin (Arazi et al. 1999). It is similar to cyclic nucleotide-gated K+ channels activated by cations and voltage. Transgenic tobacco plants have been constructed which express the gene under the control of the CaMV 35S promoter. Transgenic seedlings were able to develop at concentrations of up to 200μM Ni and tolerance was correlated with levels of the transgene product. Transgenics accumulated less Ni than did control plants at the critical concentrations for toxicity. In contrast, the transgenics accumulated more Pb and were more sensitive to it than were the non-transgenic controls. The response towards other metals was unchanged. The physiological role of this protein is still unknown, but as a facilitator for Pb uptake it could be interesting for exploitation in phytoremediation, if it can be combined with tolerance traits.

A. thaliana AtNramp3 belongs to a family of broad specificity membrane metal transporters and can complement deficiencies in manganese and iron transport in yeast (Thomine et al., 2000). It is expressed in both the roots and the aerial parts of plants, and is induced in roots by Fe starvation. Transgenic *Arabidopsis* plants over-expressing the gene under the control of the CaMV 35S promoter displayed Cd hypersensitivity and increased Fe accumulation, but the accumulation of Cd was not affected.

A. thaliana AtHMA4 encodes a P_{1B}-ATPase, a representative of a class of transporters which couple the energy from ATP hydrolysis to the translocation of positively charged substrates in membranes (Verret et al., 2004 and references therein). This protein transports Zn, Cd, Pb and Co, and is expressed in all tissues, with higher levels in roots, stems and flowers. Transgenic plants over-expressing *AtHMA4* under the control of the CaMV 35S promoter accumulated more Zn and Cd in the leaves, and displayed a higher level of tolerance to both metals. Since root-to-shoot translocation is an important factor for phytoremediation, this gene is of particular interest.

2.1.2. *Chelator systems*

Chelators have long been thought to have a role in metal accumulation. However, metallothioneins, phytochelatins and other chelators have rarely been shown to be necessary for hyper-accumulation (Yang et al., 2005). Despite contrasting evidence concerning their importance, transgenic approaches with modified expression of chelators have been extensively applied. The first target genes were mammalian metallothioneins. The human metallothionein *hMTII* was introduced into tobacco under the control of the CaMV 35S promoter with a double enhancer (de Borne et al., 1998). Transgenic plants showed increased tolerance to Cd and a modified distribution of Cd, but no changes in the total quantity of metal taken up; only 40-50% of the metal was translocated to leaves, whereas in the control non-transgenics, the proportion is as high as 80%. The distribution of other metals was not affected. As a biotechnological tool, a decrease of Cd translocation to the leaves could be an important character in the improvement of edible plants.

Histidine is a low molecular weight chelator considered to be important for the chelation of Ni (Kerkeb and Krämer, 2003). Its synthesis is limited by the enzyme ATP phosphoribosyl transferase, APRT. Transgenic *A. thaliana* plants over-expressing *St*HisG (a bacterial APRT) developed increased histidine levels and resulted in an enhanced Ni tolerance without affecting Ni accumulation (Wycisk et al., 2004).

The metabolism of sulphur plays an important role in metal response, since most chelators, including metallothioneins and phytochelatins, contain this element. *APS1* encodes an ATP sulfurylase and *Brassica juncea* has been engineered to constitutively express this gene (Pilon-Smits et al., 1999). The plants showed increased S assimilation, higher glutathione (GSH) levels and were more tolerant to Se, accumulating 2-fold higher Se levels in the shoots.

Cysteine contributes sulphur atoms to chelators, and therefore the synthesis of cysteine is a further important control point. Cysteine synthase (CSase) is the final enzyme in the biosynthetic pathway. Kawashima and colleagues (2004 and references therein) have produced tobacco plants with altered levels of this protein in the cytosol and/or chloroplasts. All transformants showed enhanced tolerance to Cd, Se and Ni, but not to Pb or Cu. In particular, the plants expressing CSase both in the cytosol and the chloroplasts had an even higher Cd tolerance, and possessed enhanced levels of Cys and GSH. The same plants also accumulated more Cd.

Phytochelatins (PCs) are important metal chelators in plants and the manipulation of their synthesis has been attempted several times, yielding contradictory results. Lee and colleagues (2003b) have created *Arabidopsis* transgenics expressing a modified gene for phytochelatin synthase *AtPCS1* under the control of its own promoter. Lines with the highest levels of expression showed an increased production of PCs in the presence of Cd, compared to wild type plants. Unexpectedly, the presence of Cd inhibited the growth of some transgenic lines, and their tolerance was therefore decreased. This Cd hypersensitivity disappeared when plants were treated with GSH, a result attributed to interference between over-expressed PCs and metal homeostasis genes.

It has been argued that enhanced PC synthesis can deplete the cellular GSH pool, and therefore several authors have attempted the engineering of GSH synthesis, either at the level of γ–glutamylcysteine synthetase or of glutathione synthase. The Pilon-Smits and Terry group has long explored these strategies in *B. juncea.* Zhu et al. (1999b) report the construction of a line over-expressing the bacterial gene *gshI* (γ–glutamylcysteine synthetase) targeted to the plastids. These plants were significantly more resistant to Cd than wild-type, produced more GSH and PCs, and Cd accumulation was increased. In the same species, engineering of glutathione synthase with the bacterial gene *gshII* over-expressed in the cytosol also generated lines with enhanced Cd tolerance, higher PC and GSH content, and increased accumulation (Zhu et al., 1999a). When these plants were tested in contaminated soil pot experiments, both classes showed increased metal accumulation. In particular the line transformed with *gshI* accumulated 2-3 times more Cr, Cu and Pb than the control wild-type (Bennett et al. 2003). The same strategy has been also applied in poplars, and as an interesting side-effect, plants engineered with γ–glutamylcysteine synthetase over-expressed in the cytosol showed increased resistance to chloroacetanilide herbicides, besides accumulating Zn (Gullner et al., 2001; Bittsánszky et al., 2004). Engineering the metabolism of GSH and PCs seems therefore to be a promising biotechnological tool for the improvement of phytoremediation capacity.

2.1.3. *Vacuolar transporters*

A further important component in the response of plant cells to heavy metal is vacuolar sequestration, mediated by tonoplast pumps. Yeast cadmium factor 1, YCF1, is a protein of the ATP-binding cassette (ABC) family and can transport Cd-GSH complexes to the vacuole. When over-expressed in transgenic *Arabidopsis* plants using four copies of the CaMV 35S enhancer, transgenics showed increased Pb and Cd tolerance, contained significantly higher levels of Cd (but not of Pb), and transported higher quantities of Cd and GSH into the vacuoles (Song et al., 2003).

2.1.4. *Metal-specific approaches*

Other transgenic approaches have addressed contamination of specific metal ions with altogether different mechanisms. The best known example is the engineering of mercury volatilisation in *Arabidopsis* and yellow poplar (*Liriodendron tulipifera*). The bacterial mer operon has been modified for expression in plant cells, resulting in the two sequences *MerA*, encoding mercuric ion reductase, and *MerB*, organomercurial lyase. Together they confer on bacteria resistance to organomercurial compounds and to ionic mercury, producing Hg^0, which is then volatilised. As a first step, *Arabidopsis* plants expressing each of the genes separately were obtained (Rugh et al., 1996; Bizily et al., 1999), and *MerAMerB* plants were selected from progeny of their hybrid. These showed resistance to organic mercury compounds and to ionic mercury and volatilised Hg from their leaves. Following this success, transgenic yellow poplar was produced (Rugh et al., 1998). In *Arabidopsis*, the protein encoded by *MerB* was targeted either to the cell wall or to the endoplasmic reticulum (Bizily et al., 2003), and in latter situation, Hg volatilization was increased. More recently, the native *merAB* operon has been introduced into tobacco chloroplasts, conferring tolerance to phenylmercuric acetate (Ruiz et al., 2003).

Selenium is another contaminant which can be volatilised. *Astragalus bisulcatus* is a resistant plant, and tolerance is attributed to the enzyme selenocysteine methyltransferase (SMT), which methylates selenocysteine (SeCys) to a non proteic amino acid, thereby decreasing the cellular concentrations of SeCys and selenomethionine, which are toxic when incorporated into proteins (LeDuc et al., 2004). This gene has been over-expressed in both *A. thaliana* and in *B. juncea*. Transgenic plants tolerated selenite, selenate and SeCys, accumulated enhanced levels of Se and volatilized more Se compared to the wild types. SeMet is not tolerated because it is not detoxified by SMT. A different approach to selenium tolerance and accumulation has used a mammalian gene for selenocysteine lyase (Pilon et al., 2003). When expressed in the cytosol and chloroplasts of *Arabidopsis*, the degradation of SeCys was increased and elemental Se was produced. Since plants expressing the protein in the cytosol were more resistant to Se than those with chloroplast expression, it was argued that accumulation of elemental Se in chloroplasts has toxic effects.

Arsenate remediation has been addressed by Dhankher and colleagues (2002) through the use of the bacterial gene for arsenate reductase (*arsC*). Plants translocate arsenate along with phosphate, whereas reduced arsenite is bound by thiol groups such as those contained in γ–glutamylcysteine. *Arabidopsis* plants have been engineered to express both arsC and γ–glutamylcysteine synthetase. To ensure expression of arsC in the aerial tissues, expression was delivered by a light-inducible promoter for a rubisco subunit. All *arsC* transgenics were hypersensitive to arsenate, since conversion of arsenate leads to a build-up of toxic arsenite. Transgenics which also contained γ–glutamylcysteine synthetase were resistant to arsenate, and hyper-accumulated arsenic. In these plants, thiol compounds act as sink for the toxic arsenite, allowing its detoxification and accumulation. *arsC* transformants in *Arabidopsis* and tobacco were also resistant to Cd (Dhankher et al., 2003). The tobacco transgenics accumulated Cd in the shoots, but the mechanism underlying this tolerance to Cd is as yet unexplained.

2.1.5. Exclusion mechanisms

Plants suitable for phytoremediation generally possess a capacity for tolerance and accumulation of heavy metals, and the previous examples have shown how these properties may be engineered. A different approach exploits the exclusion of heavy metals, useful in the agricultural context to decrease the entry of metals into the food chain. Lee et al. (1999a) utilised the bacterial gene *ZntA* which encodes a Pb, Cd and Zn pump. In transgenic *Arabidopsis* plants the protein was localised to the plasma membrane and improved tolerance to Pb and Cd, decreasing their content in shoots. Similar approaches could be useful as biotechnological tools for crop plants.

2.2. ORGANIC CONTAMINANTS

Few genetic data are available concerning the phytoremediation of organic molecules. Variability is typically documented in different species endowed with different properties, rather than in different genotypes. It is well known that the capacity to degrade organic pollutants depends on the enzyme complement, and that each species differs from others (Schwitzguébel and Vanek, 2003). Several examples of genetically engineered plants are available, in which genes encoding for specific enzymes have been transferred in order to confer a novel degradation ability. In plants, organic contaminants can be completely degraded, or they can be partially metabolized in three phases: in phase I (functionalisation) their chemical structure is modified to increase solubility and reactivity; in phase II (conjugation) the metabolites or contaminants are bound to small endogenous molecules for inactivation or targeting to storage sites; and in phase III (compartmentalisation) they are transported as conjugates to storage sites in the vacuole or cell wall (see Figure 3 in Pilon-Smits, 2005 for a summary; Schwitzguébel and Vanek, 2003). These processes have been addressed with various genetic engineering approaches (Table 1).

2.2.1. *Degradation of specific compounds*

The bacterium *Enterobacter cloacae* utilises nitrate ester explosives as source of nitrogen, via NADPH-dependent enzyme pentaerythritol tetranitrate (PETN) reductase (French et al., 1999). A modified bacterial gene, under the control of the CaMV 35S promoter, was transferred into tobacco. Tolerance to glycerol trinitrate (GTN) and trinitrotoluene (TNT) was greatly improved in transgenic plants. Wild type and transgenic plants were able to denitrify GTN to toxic glycerol dinitrate, but transgenic plants converted a higher proportion of it into the less toxic glycerol mononitrate.

A similar enzyme (nitroreductase) from the same bacterium was transferred into tobacco by Hannink et al. (2001). This enzyme utilises NAD(P)H to reduce TNT to aminodinitrotoluenes (ADNTs). Transgenic plants showed an enhanced tolerance and detoxification of TNT, although the metabolites could not be identified.

Ralstonia eutropha has the ability to use 3-chlorobenzoate as carbon source, and an important step in the pathway is the enzyme chlorocatechol 1,2-dioxygenase, encoded by the gene *cbnA*. This gene has been introduced into rice calli and plants under the control of the CaMV 35S promoter (Shimizu et al., 2002). In transgenic leaf tissues, the conversion of 3-chlorocatechol into 2-chloromuconate was time-dependent. A decrease in the level of the compound present was also observed in control plants, but this was due to adsorption as no metabolites were detected.

2.2.2. *Phase I of metabolism*

The main enzymes of Phase I are the cytochrome P450 monoxygenases, CYP450 (Morant et al., 2003). Several examples of transgenic plants modified with mammalian enzymes have been described. The human gene *2E1* was introduced in tobacco, conferring a significant increase in the metabolism of trichloroethylene (TCE) and the debromination of ethylene dibromide (Doty et al., 2000). The human genes *1A1*, *2B6* and *2C19* introduced into rice conferred resistance to herbicides belonging to six different families (Kawahigashi et al., 2005), showing high metabolic activity in the presence of chlortoluron, metolachlor and norflurazon. Recently, a plant enzyme (*76B1* from the Jerusalem artichoke *Helianthus tuberosum*) has been used to transform *Arabidopsis* and tobacco (Diderjean et al., 2002). This gene confers herbicide tolerance, and was used in both in its native form and in a construct fused to various portions of a P450 reductase gene *HTR1*. The former construct gave better results. All transgenic lines showed an increased capacity to metabolise herbicides, and the more tolerant lines had higher metabolic capacities. In the fungus *Coriolus versicolor,* the enzyme Mn peroxidase (MnP) is involved in phase I. When it was transferred into tobacco plants, the level of pentachlorophenol (PCP) present in a liquid medium was more effectively decreased by transgenic than by wild type roots (Iimura et al., 2002). Laccases are involved in the oxidation of several substrates, including polychlorynated phenols. The gene *LAC1* encodes a secretory laccase, and was isolated from cotton, *Gossypium arboreum*, and transferred to *Arabidopsis* by Wang et al. (2004). Transgenic plants showed enhanced resistance to 2,4,6-trichlorophenol and secreted laccase into the medium.

3. Summary and conclusion

Genetic analysis can provide an understanding of the mechanisms involved in phytoremediation, which otherwise can be difficult to achieve at the metabolic level using biochemical analysis. Both bottom-up and top-down approaches are useful, as exemplified by the examples described above. Classical genetic analysis (bottom-up) has shown that tolerance does not necessarily correlate with hyper-accumulation, and that tolerance and accumulation are usually inherited in a quantitative fashion. These conclusions have been largely verified by transgenic experiments. Transcriptomic analysis has demonstrated that many genes are switched on or off during the expression of tolerance and accumulation, and some have been identified using a comparative genetic approach. Gene cloning and the use of these genes for the production of transgenic plants (top-down) have confirmed some of these findings and has uncovered some novel phenotypes which could be very important for future developments.

Acknowledgements

The author thanks Elena Maestri and Marta Marmiroli for assistance in preparation of the manuscript.

References

Arazi, T., Sunkar, R., Kaplan, B. and Fromm, H., 1999, A tobacco plasma membrane calmodulin-binding transporter confers Ni^{2+} tolerance and Pb^{2+} hypersensitivity in transgenic plants, *Plant J.* **20**:171-182.

Assunçao, A. G. L., Da Costa Martins, P., De Folter, S., Vooijs, R., Schat, H., and Aarts, M. G. M., 2001, Elevated expression of metal transporter genes in three accessions of the metal hyperaccumulator *Thlaspi caerulescens*, *Plant Cell Environ.* **24**:217-226.

Assunçao, A. G. L., Ten Bookum, W. M., Nelissen, H. J. M., Vooijs, R., Schat, H., and Ernst, W. H. O., 2003, A cosegregation analysis of zinc (Zn) accumulation and Zn tolerance in the Zn hyperaccumulator *Thlaspi caerulescens*, *New Phytol.* **159**:383-390.

Becher, M., Talke, I. N., and Krämer, U., 2004, Cross-species microarray transcript profiling reveals high constitutive expression of metal homeostasis genes in shoots of the zinc hyperaccumulator *Arabidopsis halleri*, *Plant J.* **37**:251-268.

Bennett, L. E., Burkhead, J. L., Hale, K. L., Terry, N., Pilon, M., and Pilon-Smits, E. A. H., 2003, Analysis of transgenic Indian mustard plants for phytoremediation of metal-contaminated mine tailings, *J. Environ. Qual.* **32**:432-440.

Bert, V., Bonnin, I., Samitou-Laprade, P., de Laguérie, P., and Petit, D., 2002, Do *Arabidopsis halleri* from nonmetallicolous populations accumulate zinc and cadmium more effectively than those from metallicolous populations?, *New Phytol.* **155**:47-57.

Bert, V., Meerts, P., Samitou-Laprade, P., Salis, P., Gruber, W., and Verbruggen, N., 2003, Genetic basis of Cd tolerance and hyperaccumulation in *Arabidopsis halleri*, *Plant Soil* **249**:9-18.

Bittsánszky, A., Kömives, T., Gullner, G., Gyulai, G., Kiss, J., Heszky, L., Radimszky, L., and Rennenberg, H., 2005, Ability of transgenic poplars with elevated glutathione content to tolerate zinc(2+) stress, *Environ. Int.* **31**:251-254.

Bizily, S. P., Kim, T., Kandasamy, M. K., and Meagher, R. B., 2003, Subcellular targeting of methylmercury lyase enhances its specific activity for organic mercury detoxification in plants, *Plant Physiol.* **131**:463-471.

Bizily, S. P., Rugh, C. L., Summers, A. O., and Meagher, R. B., 1999, Phytoremediation of methylmercury pollution: *merB* expression in *Arabidopsis thaliana* confers resistance to organomercurials, *Proc. Natl. Acad. Sci. U.S.A.* **96**:6808-6813.

Cherian, S., and Oliveira, M. M., 2005, Transgenic plants in phytoremediation: Recent advances and new possibilities, *Environ. Sci. Technol.* **39**:9377-9390.

Courbot, M., Craciun, A., Heudiard, A., Kieffer, P., Roosens, N., Saumitou-Laprade, P., Willems, G., and Verbruggen, N., 2004, Identification of genes involved in Cd tolerance in Cd hyperaccumulation, COST Action 859, 1st WG2 Workshop – Exploiting "-omics" approaches in phytotechnologies, p. 11.

De Borne, F. D., Elmayan, T., de Roton, C., de Hys, L. and Tepfer, M., 1998, Cadmium partitioning in transgenic tobacco plants expressing a mammalian metallothionein gene, *Mol. Breeding* **4**:83-90.

Deniau, A., Pieper, B., Lindhout, P., Aarts, M., and Schat, H., 2004, QTL mapping of cadmium accumulation in *Thlaspi caerulescens*, COST Action 859, 1st WG2 Workshop – Exploiting "-omics" approaches in phytotechnologies, p. 15.

Dhankher, O. P., Li, Y., Rosen, B. P., Shi, J., Salt, D., Senecoff, J. F., Sashti, N. A., and Meagher, R. B., 2002, Engineering tolerance and hyperaccumulation of arsenic in plants by combining arsenate reductase and γ-glutamylcysteine synthetase expression, *Nat. Biotechnol.* **20**:1140-1145.

Dhankher, O. P., Sashti, N. A., Rosen, B. P., Fuhrmann, M., and Meagher, R. B., 2003, Increased cadmium tolerance and accumulation by plants expressing bacterial arsenate reductase, *New Phytol.* **159**:431-441.

Diderjean, L., Gondet, L., Perkins, R., Lau, S.-M. C., Schaller, H., O'Keefe, D. P., and Werck-Reichhart, D., 2002, Engineering herbicide metabolism in tobacco and Arabidopsis with CYP76B1, a cytochrome P450 enzyme from Jerusalem artichoke, *Plant Physiol.* **130**:179-189.

Doty, S. L., Shang, T. Q., Wilson, A. M., Tangen, J., Westergreen, A. D., Newman, L. A., Strand, S. E., and Gordon, M. P., 2000,Enhanced metabolism of halogenated hydrocarbons in transgenic plants containing mammalian cytochrome P450 2E1, *Proc. Natl. Acad. Sci. U.S.A.* **97**:6287-6291.

Dräger, D. B., Desbrosses-Fonrouge, A.-G., Krach, C., Chardonnens, A. N., Meyer, R. C., Samitou-Laprade, P., and Krämer, U., 2004, Two genes encoding Arabidopsis halleri MTP1 metal transport proteins co-segregate with zinc tolerance and account for high *MTP1* transcript levels, *Plant J.* **39**:425-439.

Ellis, D. R, Sors, T. G., Brunk, D. G., Albrecht, C., Orser, C., Lahner, B., Wood, K. V., Harris, H. H., Pickering, I. J., and Salt, D. E., 2004, Production of Se-methylselenocysteine in transgenic plants expressing selenocysteine methyltransferase, *BMC Plant Biol.* **4**:1.

French, C. E., Rosser, S. J., Davies, G. J., Nicklin, S., and Bruce, N. C., 1999, Biodegradation of explosives by transgenic plants expressing pentaerythritol tetranitrate reductase, *Nat. Biotechnol.* **17**:491-494.

Frérot, H., Petit, C., Lefèbvre, C., Gruber, W., Collin, C., and Escarré, J., 2003, Zinc and cadmium accumulation in controlled crosses between metallicolous and nonmetallicolous populations of *Thlaspi caerulescens* (Brassicaceae), *New Phytol.* **157**:643-648.

Gullner, G., Körnives, T., and Rennenberg, H., 2001, Enhanced tolerance of transgenic poplar plants overexpressing γ-glutamylcysteine synthetase towards chloroacetanilide herbicides, *J. Exp. Bot.* **52**:971-979.

Hannink, N., Rosser, S. J., French, C. E., Basran, A., Murray, J. A. H., Nicklin, S., and Bruce N. C., 2001, Phytodetoxification of TNT by transgenic plants expressing a bacterial nitroreductase, *Nat. Biotechnol.* **19**:1168-1172.

Kawahigashi, H., Hirose, S., Inui, H., Ohkawa, H., and Ohkawa, Y., 2005, Enhanced herbicide cross-tolerance in transgenic rice plants co-expressing human CYP1A1, CYP2B6, and CYP2C19, *Plant Sci.* **168**:773-781.

Kawashima, C., G., Noji, M., Nakamura, M., Ogra, Y., Suzuki, K. T., and Saito, K., 2004, Heavy metal tolerance of transgenic tobacco plants over-expressing cysteine synthase, *Biotechnol. Lett.* **26**:153-157.

Kerkeb, L., Krämer, U., 2003, The role of free histidine in xylem loading of Nickel in *Alyssum lesbiacum* and *Brassica juncea*, *Plant Physiol.* **131**: 716-724.

LeDuc, D. L., Tarun, A. S., Montes-Bayon, M., Meija, J., Malit, M. F., Wu, C. P., AbdelSamie, M., Chiang, C.-Y., Tagmount, A., deSouza, M., Neuhierl, B., Böck, A., Caruso, J., and Terry, N., 2004, Overexpression of selenocysteine methyltransferase in Arabidopsis and Indian mustard increaes Selenium tolerance and accumulation, *Plant Physiol.* **135**:377-383.

Lee, J., Bae, H., Jeong, J., Lee, J.-Y., Yang, Y.-Y., Hwang, I., Martinoia, E., and Lee, Y., 2003a, Functional expression of a bacterial heavy metal transporter in Arabidopsis enhances resistance to and decreases uptake of heavy metals, *Plant Physiol.* **133**:589-596.

Lee, S., Moon, J. S., Ko, T.-S., Petros, D., Goldsbrough, P. B., and Korban, S. S., 2003b, Overexpression of Arabidopsis phytochelatin synthase paradoxically leads to hypersensitivity to cadmium stress, *Plant Physiol.* **131**:656-663.

Macnair, M. R., 2002, Within and between population genetic variation for zinc accumulation in *Arabidopsis halleri*, *New Phytol.* **155**:59-66.

Meerts, P., and Van Isacker, N., 1997, Heavy metal tolerance and accumulation in metallicolous and non-metallicolous populations of *Thlaspi caerulescens* from continental Europe, *Plant Ecol.* **133**:221-231.

Morant, M., Bak, S., Lindberg Moller, B., Werck-Reichhart, D., 2003, Plant cytochromes P450: tools for pharmacology, plant protection and phytoremediation, *Curr. Opin. Biotechnol.* **14**:151-162.

Paterson, A. H., Lander, E. S., Hewitt, J. D., Peterson, S., Lincoln, S. E., Tanksley, S. D., 1988, Resolution of quantitative traits into Mendelian factors by using a complete linkage map of restriction fragment length polymorphisms, *Nature* **335**:721–726.

Pilon, M., Pwen, J. D., Garifullina, G. F., Kurihara, T., Mihara, H., Esaki, N., and Pilon-Smits, E. A. H., 2003, Enhanced selenium tolerance and accumulation in transgenic Arabidopsis expressing a mouse selenocysteine lyase, *Plant Physiol.* **131**:1250-1257.

Pilon-Smits, E. A. H., Hwang, S., Lytle, C. M., Zhu, Y., Tai, J. C., Bravo, R. C., Chen, Y., Leustek, T., and Terry, N. 1999, Overexpression of ATP sulfurylase in Indian mustard leads to increased selenate uptake, reduction, and tolerance, *Plant Physiol.* **119**:123-132.

Pilon-Smits, E., 2005, Phytoremediation, *Annu. Rev. Plant Biol.* **56**:15-39.

Rugh, C. L., Wilde, H. D., Stack, N. M., Marin Thompson, D., Summers, A. O., and Meagher, R. B., 1996, Mercuric ion reduction and resistance in transgenic *Arabidopsis thaliana* plants expressing a modified bacterial *merA* gene, *Proc. Natl. Acad. Sci. U.S.A.* **93**:3182-3187.

Rugh, C. L., Senecoff, J F., Meagher, R. B., and Merkle, S. A., 1998, Development of transgenic yellow poplar for mercury phytoremediation, *Nat. Biotechnol.* **16**:925-928.

Ruiz, O. N., Hussein, H. S., Terry, N., and Daniell, H., 2003, Phytoremediation of organomercurial compounds via chloroplast genetic engineering, *Plant Physiol.* **132**:1344-1352.

Schwitzguébel, J.-P., and Vanek, T., 2003, Some fundamental advances for xenobiotic chemicals, in: *Phytoremediation: Transformation and control of contaminants,* S. C. McCutcheon, J. L. Schnoor, eds., John Wiley & Sons, Inc., Hoboken, pp. 123-157.

Shimizu, M., Kimura, T., Koyama, T., Suzuki, K., Ogawa, N., Miyashita, K., Sakka, K., and Ohmiya, K., 2002, Molecular breeding of transgenic rice plants expressing a bacterial chlorocatechol dioxygenase gene, *Appl. Environ. Microbiol.* **68**:4061-4066.

Song, W.-Y., Sohn, E. J., Martinoia, E., Lee, Y. J., Yang, Y.-Y., Jasinski, M., Forestier, C., Hwang, I., and Lee, Y., 2003, Engineering tolerance and accumulation of lead and cadmium in transgenic plants, *Nat. Biotechnol.* **8**:914-919.

Thomine, S., Wang, R., Ward, J. M., Crawford, N. M., and Schroeder, J. I., 2000, Cadmium and iron transport by members of a plant metal transporter family in *Arabidopsis* with homology to *Nramp* genes, *Proc. Natl. Acad. Sci. U.S.A.* **97**:4991-4996.

Van de Mortel, J., Rigola, D., Talukdar, S., Fiers, M., Pieper, B., Moller, B., Schat, H., and Aarts, M., 2004, Comparative genomics of Zn accumulation in *Thlaspi caerulescens* and *Arabidopsis thaliana*, COST Action 859, 1st WG2 Workshop – Exploiting "-omics" approaches in phytotechnologies, p. 8.

Verret, F., Gravot, A., Auroy, P., Leonhardt, N., David, P., Nussaume, L., Vavasseur, A., and Richaud, P., 2004, Overexpression of AtHMA4 enhances root-to-shoot translocation of zinc and cadmium and plant metal tolerance, *FEBS Lett.* **576**:306-312.

Via, S., Gomulkiewicz, R., De Jong, G., Scheiner, S. M., Schlichting, C. D., and Van Tienderen, P. H., 1995, Adaptive phenotypic plasticity: consensus and controversy, *TREE* **10**:212-217.

Wang, G.-D., Li, Q.-J., Luo, B., and Chen X.-Y., 2004, *Ex planta* phytoremediation of trichlorophenol and phenolic allelochemicals via an engineered secretory laccase, *Nat. Biotechnol.* **7**:893-897.

Weber, M., Harada, E., Vess, C., v. Roepenack-Lahaye, E., and Clemens, S., 2004, Comparative microarray analysis of *Arabidopsis thaliana* and *Arabidopsis halleri* roots identifies nicotianamine synthase, a ZIP transporter and other genes as potential metal hyperaccumulation factors, *Plant J.* **37**:269-281.

Willems, G., and Samitou-Laprade, P., 2004, Genetic characterisation and mapping of Quantitative Trait Loci involved in zinc tolerance and accumulation in *Arabidopsis halleri* subsp. *halleri*, COST Action 859, 1st WG2 Workshop – Exploiting "-omics" approaches in phytotechnologies, p. 45.

Wycisk, K., Kim, E. J., Schroeder, J. I., and Krämer, U., 2004, Enhancing the first enzymatic step in the histidine biosynthesis pathway increases the free histidine pool and nickel tolerance in *Arabidopsis thaliana*, *FEBS Lett.* **578**:128-134.

Yang, X., Feng, Y., He, Z. L., and Stoffella, P. J., 2005, Molecular mechanisms of heavy metal hyperaccumulation and phytoremediation, *J. Trace Elem. Med. Biol.* **18**:339-353.

Zha, H. G., Jiang, R. F., Zhao, F. J., Vooijs, R., Schat, H., Barker, J. H. A., and McGrath, S. P., 2004, Co-segregation analysis of cadmium and zinc accumulation in *Thlaspi caerulescens* interecotypic crosses, *New Phytol.* **163**:299-312.

Zhu, Y. L., Pilon-Smits, E. A. H., Jouanin, L., and Terry, N., 1999a, Overexpression of glutathione synthetase in Indian mustard enhances cadmium accumulation and tolerance, *Plant Physiol.* **119**:73-79.

Zhu, Y. L., Pilon-Smits, E. A. H., Tarun, A. S., Weber, S. U., Jouanin, L., and Terry, N., 1999b, Cadmium tolerance and accumulation in Indian mustard is enhanced by overexpressing γ-glutamylcysteine synthetase, *Plant Physiol.* **121**:1169-1177.

PART 3

FEASIBILITY STUDIES AND PRACTICAL IMPLEMENTATION

DECHLORINATION OF CHLORINATED HYDROCARBONS BY ZERO-VALENT IRON NANO-PARTICLES

M. PUPEZA*, M. CERNIK**, M. GRECO*
** Golder Associates Europe Ltd., Via Antonio Banfo 43, Torino, Italy, mpupeza@golder.it*
***AQUATEST a.s., Geologická 4, Praha, Czech Republic, cernik@aquatest.cz*

*To whom correspondence should be addressed.

Abstract- Nano-particle zero-valent iron has been used for several years by Golder Associates in the United States to reduce chlorinated hydrocarbons. Its first application (in the form of a pilot test) in Europe was successfully performed at a large industrial facility in the Czech Republic. The test site, located on alluvial terraces of the Elbe River, is among the oldest chemical production facilities in Europe, where a large array of chemicals (from fertilizers and detergents to chlorinated solvents) was manufactured over several decades. Lack of awareness regarding environmental protection and continuous neglect have resulted in the release to the sub-surface of significant quantities of chlorinated solvents, now present both as free product and in the dissolved phase in the aquifer underlying the site, and presently endangering the water quality of the Elbe River. Groundwater chemistry has been changed by this human activity, and is characterised by the presence of about 8mg/l chlorinated solvents and up to 500mg/l sulphates. Preliminary tests involved spike and bench tests as a means to calibrate the field test. Several concentrations of iron were experimented with during preliminary tests. The pilot test consisted of the injection of iron nano-particles into one well, and the monitoring of hydraulic and chemical parameters over several months, both in the injection well and in two monitoring wells located up- and down-gradient of the injection site. The test results showed a rapid decrease in solvent concentrations in the first week after injection, and concentrations remained low thereafter. Iron particles were detected in the down-gradient monitoring well, proving that they can move with groundwater, thus enhancing their spatial effectiveness.

N. Marmiroli et al. (eds.), Advanced Science and Technology for Biological Decontamination of Sites Affected by Chemical and Radiological Nuclear Agents, 111–118.

Keywords: Zero-valent iron; nano-particles; chlorinated hydrocarbons; dehalogenation; remediation

1. Introduction

This article presents the approach, methodology and results obtained during a pilot test experimenting with the use of zero-valent iron (ZVI) particles at a site contaminated with chlorinated hydrocarbons. We describe spike and bench tests, the installation of two monitoring wells and the testing of the local hydro-geological parameters, the collection of samples, the analysis of baseline conditions prior to the test, the injection of a nano-particles slurry, and the short- and long-term post-injection monitoring.

2. Theoretical background

Present technologies used for the remediation of groundwater affected by pollutants such as chlorinated hydrocarbons include “pump and treat”, *in situ* chemical oxidation, and enhanced bioremediation. The latter is limited by the conditions of the site and the time frame required in order to achieve clean-up. More than a decade ago, permeable reactive iron barriers (less than 500m^2/kg of reactive surface) were shown to achieve dechlorination by means of chemical reduction. These have performed well, but the capital cost of their installation and the disturbance they create on site are significant disadvantages to their widespread use. Consequently, steps have been made to optimize the process by reducing the size of the ZVI particles; those currently produced by Golder Associates are about 50nm in diameter (the size of a bacterial cell), thereby increasing their reactive surface up to 30,000m^2/kg, while at the same time allowing passive transport of the particles away from the injection point.

The chemical process is simple and straightforward. Under highly anaerobic conditions water is reduced by iron:

$$Fe^0 + 2H_2O \rightarrow Fe^{2+} + H_2 + 2OH$$

Similarly, the abiotic dehalogenation of chlorinated hydrocarbons such as, for example, trichloroethylene (TCE) to ethene and chloride occurs by the transfer of electrons resulting from the oxidation of iron:

$$3Fe^0 + C_2HCl_3 + 3H^+ \rightarrow 3Fe_2+ + C_2H_4 + 3Cl^-$$

The first pilot test of ZVI decontamination conducted by Golder Associates in Europe in 2003 is presented below.

3. General site conditions

The site is located in the north-western part of the Czech Republic in an area where heavy and chemical industries have operated over more than a century. The Slovay plant is probably the oldest chemical manufacturing facility in Europe (founded in about 1868). Production has been more or less continuous ever since, and although it is presently being down-sized, the facility is still in operation. A large variety of chemicals, from chlorinated solvents to sulphate- and nitrate-based chemicals, has been produced over the past 40 years.

4. Conceptual site model

The site is located on the west bank of the Labe (Elbe) River, on terraces of fluvial deposits that exhibit irregular layering of variable grain size and hydraulic conductivity. The geology and hydro-geology of the area is complicated. Designing and conducting a pilot test needed to be preceded by a complex analysis of geological, hydro-geological and geochemical data, in order to develop a good understanding of local site conditions and the creation of a sound conceptual site model. Site investigations revealed the impact of chlorinated solvents found in the sub-surface in both free- and dissolved phase over widespread areas, referred to as follows:

Plume No. 1 (areas 1, 3 and 6 in Figure 1) is located in low permeable silty clay deposits to 26m below ground surface (bgs). The groundwater table is 6m bgs and the hydraulic conductivity, k, is 3×10^{-7}m/s. Concentrations of chlorinated solvents vary from 10mg/l to free phase.

Plume No. 2 (area 2) is located in low permeable, partially fractured loess ($k = 8 \times 10^{-8}$m/s) ranging to 4m bgs, underlain as far as 9m bgs by sand and gravel ($k = 7 \times 10^{-5}$m/s); below 9m bgs is clay. The groundwater table is at 4m bgs. Contamination was found throughout the whole profile, in concentrations between 5mg/l to free phase.

Plume No. 3 (areas 4 and 5) is located in low permeable, partially fractured clay from 0 – 10m bgs ($k = 4 \times 10^{-8}$ m/s), over a layer of sand and gravel ($k = 5 \times 10^{-5}$ m/s) down to 18m bgs, and underlain by clay. The groundwater table is at 10m bgs in the sand and gravel. In areas where this layer is absent, only perched groundwater occurs, at depths of between 2 and 12m bgs. Contamination was found throughout the whole profile, in concentrations between 15mg/l to free phase, with most of the free phase found in the upper clay layer.

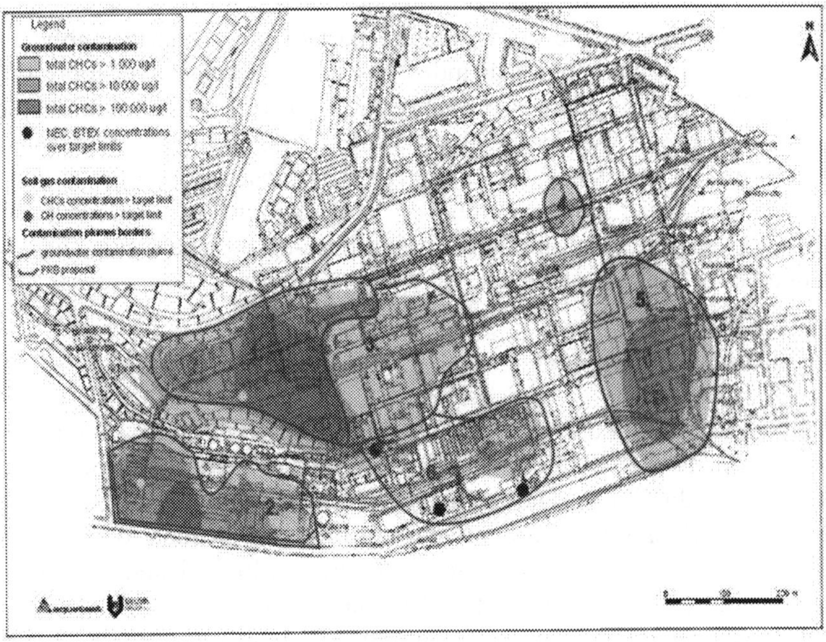

Figure 1.

Two test areas were selected, one in Plume No. 1 in an area underlain by silty clay with a hydraulic conductivity of 3 x 10^{-7} m/s; and the second in Plume No. 3 in sand and gravel deposits with a hydraulic conductivity of 5 x 10^{-5} m/s.

5. Test design

Spike and bench-scale tests were performed in order to size the pilot test. For this purpose, soil samples of about 10kg each and groundwater samples (about 5kg each) were retrieved/collected both during and after the installation of the test wells in, respectively, sealed plastic bags and glass containers, and these were preserved under refrigeration. The samples were analysed for content of chlorinated solvents, sulphates, nitrates and metals (baseline analysis).

Various concentrations of nano-scale ZVI were run through the test columns to determine the optimal concentration to be employed during the field test. Groundwater samples were collected at regular intervals and analysed for the same contaminants as the baseline samples. In addition,

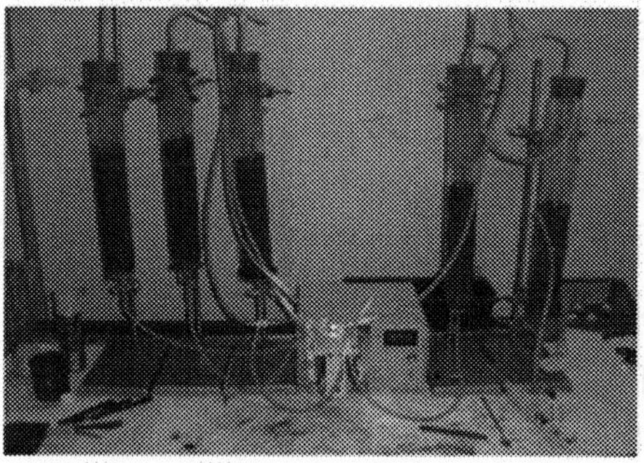

Bench test columns

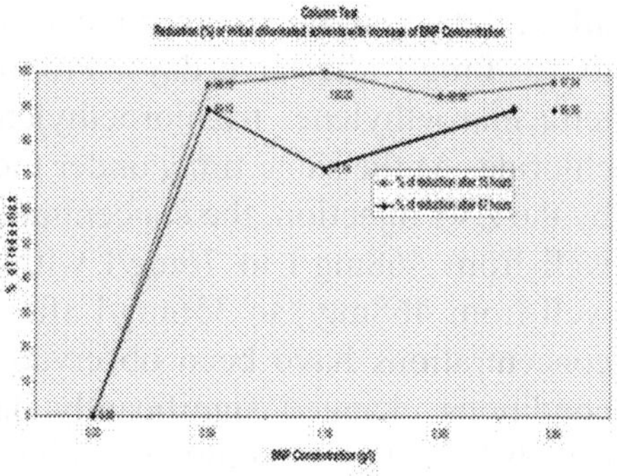

Reduction of solvents concentration with increase of iron concentrations

Figure 2.

dissolved oxygen (DO), oxidation reduction potential (ORP) and Eh/pH were continuously monitored.

The results of these tests were used to size the pilot tests. For reasons of space, only the test conducted in Plume Area No. 3 will be discussed in detail.

6. Pilot test

The pilot test site consisted of three monitoring wells fully penetrating the aquifer. These were located at 3m intervals along the groundwater flow direction. The middle well was used as the injection well, while the up- and down-gradient wells served as monitoring wells. Prior to injection, groundwater samples were collected from the three wells and analysed to obtain baseline values. A total mass of 5kg of ZVI nano-particles was mixed with drinking water and injected over 24 hours. DO, ORP, pH, temperature and conductance, as well as groundwater level were monitored continuously in all three wells by means of electronic data loggers, both during injection and for one month after injection. In addition, groundwater samples were collected from the three wells and analysed according to a pre-established schedule for one year after the test.

7. Results

The effect of nano-scale ZVI injection and the related results agree well with previous laboratory and field tests undertaken by Golder Associates. Total chlorinated hydrocarbons decreased in the injection well from 7.076mg/l to 0.913mg/l within 24 hours of ZVI injection and to 0.410mg/l after 28 days; in the down-gradient well, the reduction was from 6.597mg/l to 1.640mg/l after five days. Proportional reductions were monitored for tetrachloroethylene, trichloroethylene, cis 1,2-dichloroethylene and vinyl chloride. At the same time, under the anaerobic reducing conditions created by the ZVI injection, the concentration of sulfates decreased in the injection well from 468mg/l to 76mg/l within 24 hours; and in the down-gradient well from 468mg/l to 356mg/l after five days. Similar decreases in nitrate concentrations have been observed. Pre-injection baseline redox and DO conditions changed significantly during the injection, demonstrating the establishment of a strongly reducing environment.

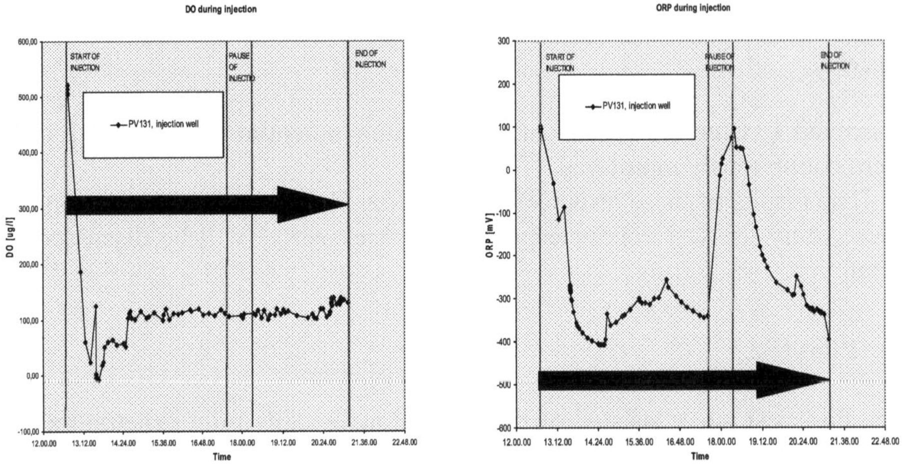

Figure 3.

ZVI particles were found in the groundwater sample collected from the down-gradient monitoring well two weeks after the injection, showing that ZVI moves with the groundwater. As far as we know, this is the first time movement of ZVI in the sub-surface has been observed.

A continuous decrease in the conductivity, and in the concentrations of chlorinated hydrocarbons and sulphates was seen during the first month post-injection. The concentrations remained low for the next five months, increasing within one year of the injection of ZVI to roughly the level of the pre-test values. This increase is due to the small volume of ZVI injected, combined with the highly affected area where the test was performed. Contaminated groundwater up-gradient of the test area has entered the area, exhausting the injected ZVI mass.

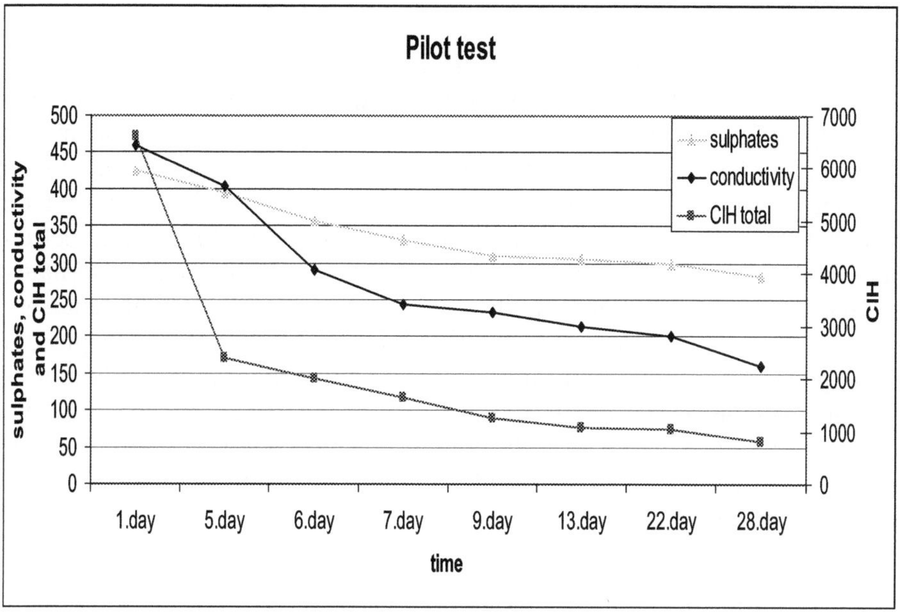

Figure 4.

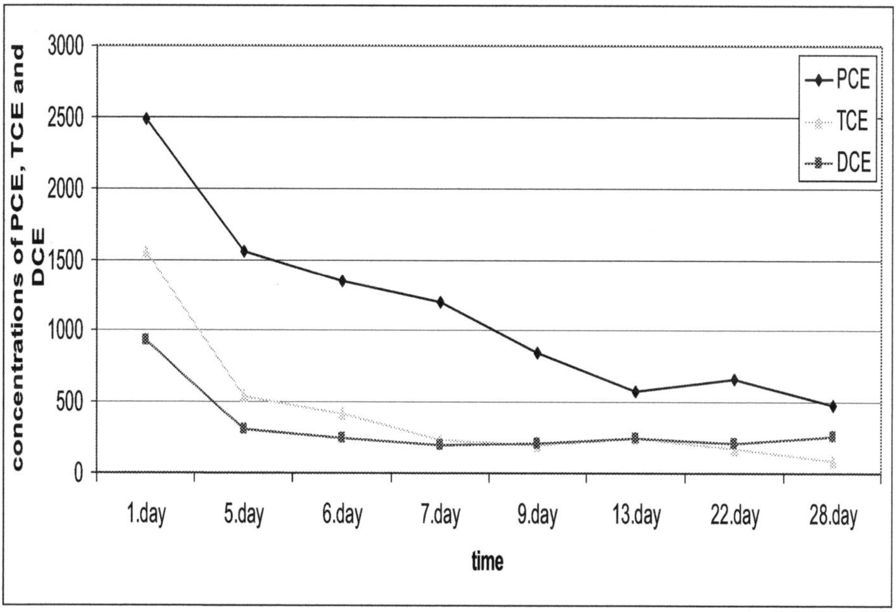

Figure 5.

8. Conclusions

The first ZVI nano-particle injection test performed by Golder Associates in Europe, in cooperation with Aquatest, yielded very positive results and new proprietary knowledge regarding the technology. It has paved the way to subsequent tests and designs for full scale remediation activities. Concentrations of chlorinated solvents decreased dramatically during the first hours/days after the injection and remained low for a long period, consistent with the mass balance between ZVI and chlorinated hydrocarbons in and around the test area. Concentrations of sulphates and nitrates decreased in parallel with the solvents, although this took place at the expense of a diminished effect of the ZVI injection on the reduction of solvents. As far as we know, the test demonstrated for the first time that ZVI nano-particles move away from their injection point with groundwater, suggesting that ZVI can be used for the treatment of larger areas of contaminated aquifers.

Acknowledgements

Michael Pupeza, CPG is a senior practice leader and Principal of Golder Associates, presently based in Turin; Dr. Miroslav Cernik is a senior consultant with Aquatest in Prague and researcher with the University of Liberec the Czech Republic, Marco Greco is an environmental engineer operating out of Golder's Turin office

REMEDIATION ALTERNATIVES ON HIGH TOXICITY AND DANGEROUS SITES

DAVID L . RUSSELL
Global Environmental Operations, Inc.
4642 Warrior Trail, SW., Lilburn, GA, 30047, USA *.

Abstract- Not all sites can be remediated by phytoremediation techniques. Sometimes, more conventional methods have to be used. This paper explores some of the decision-making process, and the techniques and costs involved in dealing with high toxicity and dangerous sites.

Keywords: High Toxicity, Radioactive, Chemical Agents, Remediation, Costs

1. Introduction

Remediation is a process of restoration. In order to conduct remedial activities, we must first understand both the process which generated the contaminants, the contaminant chemistry, the soil geology, the site hydrology, and the nature and kind of interactions between the soil, water, and the contaminants. In this brief exposition, we'll discuss a couple of different types of sites and see the limits of remediation technology.

1.1. COMMON ELEMENTS IN REMEDIATION

The contamination and the solutions in remediation have three common elements. The first is the source. If the source is contained, the contamination will stop spreading. The second element is the vector. The vector can be gravity, groundwater, air, surface water, wind, fire, or whatever moves the contamination off or away from the source. The third is the receptor. The receptor is the most critical element because it is where the harm comes in or where the problem is destroyed. Sometimes, however, a single source will have multiple vectors and multiple receptors. The Figure 1 indicates this.

N. Marmiroli et al. (eds.), Advanced Science and Technology for Biological Decontamination of Sites Affected by Chemical and Radiological Nuclear Agents, 119–138.

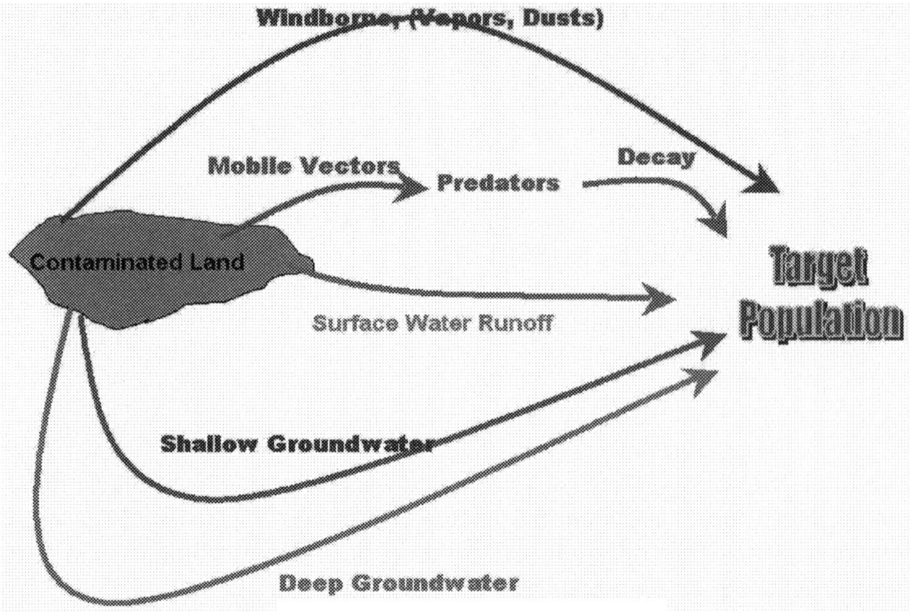

Figure 1. Source vector and target.

A brief example helps: If I have a jar containing gasoline it is a potential explosion hazard, or is it? If the jar is sealed vapor tight, where's the hazard? Now take the cap off the jar? Is there a hazard? If so, to whom? The air blowing across the jar and the vapor pressure of the liquid will cause it to evaporate. Again, no problem — maybe or maybe not. However, if we are in a confined space with an source of ignition, there may be an explosion. Similarly, if the container falls off the shelf and releases the gas, there is an additional hazard which may lead to fire or explosion.

The cylinder of an automobile is a confined space and we have explosions all the time making our vehicles run, but they are under control where and when we want them and the receptor and the vectors are well understood.

However if we have a confined space not designed for the explosion, and have a vapor and a source of ignition, we may have flying debris or fire. Again, if nobody is around this is not a problem. If you are in the building when the gasoline ignites or the vapor explodes you may have a life-threatening problem. It all depends upon the source and the vector and the receptor. Break the chain between the three and the harm is eliminated or controlled.

1.2. CONSIDERING THE CHEMICALS PRESENT

On high toxicity sites even with a very small amount of material escaping from the site, may have extreme consequences. Military Agents are high toxicity compounds which are designed to be lethal in very very small quantities. They are also extremely stable and hard to destroy. A number of industrial chemicals, particularly pesticides also fall into this category. Chlorinated solvents, PCB's and some pesticides, herbicides, and biocides fall into this category.

Some of the older pesticides and some of the nerve agents are in the same family, neurotoxins and endocrine disruptors, and have been placed on a restricted or hazardous chemical list by various environmental agencies. A few of these compounds have heavy metals attached to them. Of the heavy metals, the most common and difficult include lead, arsenic, chromium and mercury. Some of the most highly toxic compounds include the organophosphides, such as VX (O-ETHYL-S-(2-IISOPRO-PYLAMINOETHYL) METHYL PHOSPHONOTHIOLATE) and Parathion (0,0-diethyl-0-p-introphenylphosphohorothioate), and some of the highly chlorinated organic compounds such as the polychlorinated biphenyls.

Note that there is a difference in toxicity and effect. While VX and Parathion can kill quickly, PCB's have a different route and are primarily more deadly to the higher organisms, via longer term attack of the liver and kidneys. DDT is a banned pesticide, but is not on the list of high toxicity compounds; 2,4-Dichlorophenoxyacetic Acid (2,4,D), the active ingredient in Agent Orange is considered one of the high toxicity compounds. The scope of the mechanism of the insult to the body or the effects of the compound are not part of this presentation, but toxicity levels are because they influence the remediation of the site.

Fortunately for us, we have not seen any recent situations where there was massive use of chemical warfare agents outside of the Kurdish regions of Iraq. The noticeable exceptions to this are the sites left over from WWI, and some training sites used by the various military establishments in the US, and Eastern Europe. There may be many other sites, but I'm just not aware of them.

Table 1. Agents, Chemicals, and Toxicity.

Compound	**Class or Principal Use**	Human **Toxic Effects**	**Environmental Toxic Effects**
VX O-ETHYL-S-(2-IISOPROPYLAMINOETHYL) METHYL PHOSPHONOTHIOLATE	Chemical Nerve Agent	LD50 10 mg/ 70 Kg Man (Skin) or 0.14 mg/Kg	Highly Persistent, does not biodegrade. Half life over of 10^4 hours[2]
MUSTARD Bis(2-chloroethyl) sulfide	Vesicant -- Chemical Blister Agent	100 mg/ Kg (skin) 1500 mg-min/M3 (air)	A blister agent which causes painful blist-ers, and attacks the mucous membranes.
Lewisite (L, 2Çchloroviny ldichloroarsine, 2-chlorovinyl arsonous dichloride)	Vesicant – blister agent with Arsenic	30-50 mg/Kg (skin)	Highly persistent because of Arsenic in formulation
DDT (dichlorodiphenyltrichloroethan e)	Pesticide	Well above 278 mg/Kg[3]. - None identified EPA identified as of Moderate Toxicity 2000-5000 mg/Kg rat -skin LD50	Will naturally decay to DDE and then stop all further degradation
24D 2,4-Dichlorophenoxyacetic acid	Herbicide (Agent Orange)	639->5000 mg/Kg in mammals	30- 60 days in soil
Methyl Parathion	Pesticide	LC50 (mg/L/4 hrs) = 0.119	Half life in soil >20 days
PERC (Tetrachloroethylene)	Industrial Solvent	Prolonged Exposure above 1500 mg/ M3	Surface of vegetables 10-30 days

[2] Sources for Table 1 include: Kingery & Allen's article on "The Environmental Fate of Organophosphorous Nerve Agents, a Review" which appeared in Toxilogical and Environmental Chemistry, V 47, pp 155-184, Overseas Publishers Association , Amsterdam , 1994

1.2.1. *Commonalities and defining factors*

High Hazard sites often contain highly toxic compounds. For the potential application of phytoremediation to one of these sites, toxicity to the plant life, and human hazards are the determining factors. The two can be markedly different, both in kind and effect. The approach to a remedial candidate which contains an heavy metal, an endocrine disruptor, or an herbicide, or a nuclear compound are all significantly different. We have to classify the risk somehow, and whether the risk is to humans or plants, or the environment at large, and depending upon the site uses, we often set the protection standards for children, on the assumption that they have the lowest body mass, and have the greatest potential susceptibility to any compound which might be harmful.

1.2.2. *First screening level*

The first screen is toxicity and environmental effects of the compounds. Table 1 shows the relative value of toxicities for several different types of industrial compounds and for several chemical warfare agents, and compares them to some common industrial pesticides and chemicals.

The list is far from complete or comprehensive. In most instances, the data on toxicity can be obtained by locating an appropriate Material Safety Data Sheet for the compound. A quick web search can identify the toxic levels for most compounds. The site must be evaluated with respect to the levels of compounds, their toxicity and the likelihood of deliberate or accidental contact between workers on the site and the chemicals. A suitable level of protective clothing must be selected for all workers.

1.2.3. *Second screening level*

The second screen must be the soil/chemical interactions. This is necessarily more complex a screen as it involves both the soil type and the mechanism of bonding between the chemical and the soil particles. Broadly speaking, you have two separate types of interactions -- chemical bonding and surface bonding. Chemical bonding involves a reaction such as are found in certain types of clay soils where there is an chemical interaction between the soil and the target chemical. These are often characterized by the ability of the soil to act as an ion exchange agent or natural zeolite. Many clays and other minerals have this ability.

---

[3] A two year old child accidentally ingested DDT/Kerosene mixture equivalent to 278 mg/Kg and died. Workers have experienced substantially higher concentrations without ill effects.

When we look at surface bonding interactions, there are two different types - adsorption and absorption. Adsorption is an interaction of a compound on the surface of a soil particle due primarily to the surface charge (usually negative) of the soil particle. Adsorption can often be reversed by simply changing the pH of the soil with an acid wash to release the compound. By comparison, the absorption is an internal phenomenon, primarily due to the organic content of the soil. It is still a surface phenomenon because it is easily reversed, but the difference is that it most often takes place inside the soil particle. For comparative purposes, think of the differences between sand and activated carbon. The carbon holds compounds on the inside, while the sand will hold compounds on its surface. The sand can be washed to restore its' original properties, the carbon must be thermally or chemically regenerated. Compounds like Arsenicals can often be removed from sands by simply adjusting the pH while PCB's and their like tend to be absorbed in the organic fraction of the soils, and have to be thermally regenerated or burned out.

1.2.4. *Third screening level*

The third set of differentiating characteristics are the physical parameters of the soil. These are determined by a combination of routine ASTM tests for soil grain particle size, soil density, plasticity, bearing strength, compaction, organic content, void ratio, and measured in-situ hydraulic properties such as hydraulic conductivity, and permeability. The physical measurements help classify the soils for such things as ease of transport, capacity to generate dusts, and ease of handling. The second set of tests measure the properties of the soil to transmit water and other fluids.

With these tests we can not only differentiate sand, silts, and clays, but we establish a uniform basis for determining which fractions may be important in our remediation and which fractions contain the mass of the target compound. We can also gain some idea of the properties of the soils for material handling, and, most importantly, the ability of the compound to move freely through the soil with water.

The hydrological characterization of the soil may have already been performed as a part of a preliminary remedial investigation, but much of the information has been developed with regard to defining the extent of the contamination and not with remedial solutions in mind, and not all of it may be suitable for your purposes.

For example, if an excavation needs to be performed, ground water is always an issue. In order to remove the most concentrated mass of soils, one may have to dig below the groundwater table. If you are continually dealing with groundwater in an excavation, more often than not, your excavation bottom below the surface may be a highly liquid soup of groundwater and soil. If this occurs, the excavation, dewatering and disposal costs are greatly increased. Dewatering may be required.

1.2.5. *Fourth screening level - groundwater*

If you have to dewater the excavation on a contaminated site, you will have to deal with and treat the contaminated groundwater, not only identifying and managing a suitable treatment technology, but also finding a way to deal with the volume of the now treated formerly contaminated groundwater, water which the regulatory agency won't let you discharge back into the ground at all or into a river until it has been exhaustively tested. Any you will have to deal with the volume of contaminated treatment agents and materials which contain that which you have removed from the groundwater. This may include muds, sludge, chemicals, activated carbon, and oils. It is important to note that these may be the most difficult and expensive elements to handle and to dispose.

1.2.6. *Fifth screening level*

The fifth set of screening criteria are the technologies for ultimate destruction and site cleanup. The screening criteria are most often based upon the destruction removal efficiency and the end products of that activity plus a judgment factor on the ease of application and cost of application of that technology. The appropriate technologies must consider both the end product and the logistical elements and costs for implementing. The following Table 2 may be of some assistance in developing that screening. It's a sort of mini-RIFS process you can use for initial technology screening.

1.3. SCREENING TECHNOLOGY MATRIX

A recently developed approach by the EPA is compatible with the idea of screening technologies. EPA has used their website http://www.clu-in.org to describe various remediation technologies and the state of development and potential applications to existing contaminants. A sample of their website is shown in Figure 2 (EPA's Table 3.2 from their clu-in.org website). Their rankings include assessment of the development of the technology and applicability to specific contaminants but do not consider the soil and geological conditions.

Table 2 is a slightly different view of the applicability of the remediation process. It is broadly divided into organic and inorganic contaminants, and considers the ease of applying a specific technology to a soil type. The categories and assumptions in Table 2 are very broad, and many assumptions and qualifications may not be not at first apparent. However, it is a place to start.

Embedded in the Table 2 is my specific view of the universe of remediation technologies and their limitations, and that's one person's view only! For example, Incineration is an excellent technology but because of the cost of excavation and transportation, it may or may not be suitable for a specific site. Similarly, thermal methods are not generally applicable for treatment of water because they are uneconomic and difficult to sustain. Where soils may have very high water contents, incineration gets a lower ranking. Bioremediation gets a lower ranking in tight soils and highly impervious clays. Part of the reason for that is that water does not move through the soils and one has to get the contaminant in contact with the bacteria in order to conduct remediation.

Which ever set of tables you use to develop the initial information, you have to address both the technology and the soils. They are a unit, and consideration of one without the other is a sure recipe for significant cost over-runs and potential failure of the remediation scheme.

2. Important questions

There is a best technology for each particular combination of soil, contaminant and environment. One size or one type of technology does not fit all applications.

When we are dealing with a high toxicity site, there are environmental exposure criteria as well as the questions about the soils and the technology, and we must be able to answer two very important questions:

9. What is the end product of the technology to be applied, and
10. Where is Away?

The first question causes you to think about what or how you are going to do what you want to do, and the second forces you to consider the final disposition of your activities.

2.1. TECHNOLOGY VS. AWAY

The technology question can be answered quite readily, but the question about "Away" while not at first apparent is often the most important point of the discussion. On almost every process flow sheet or process development schematic, there is a little truck down at the lower right hand corner for waste disposal. This is the "Away". Years ago, we assumed that the wastes would be taken care of and disposed properly. But because they

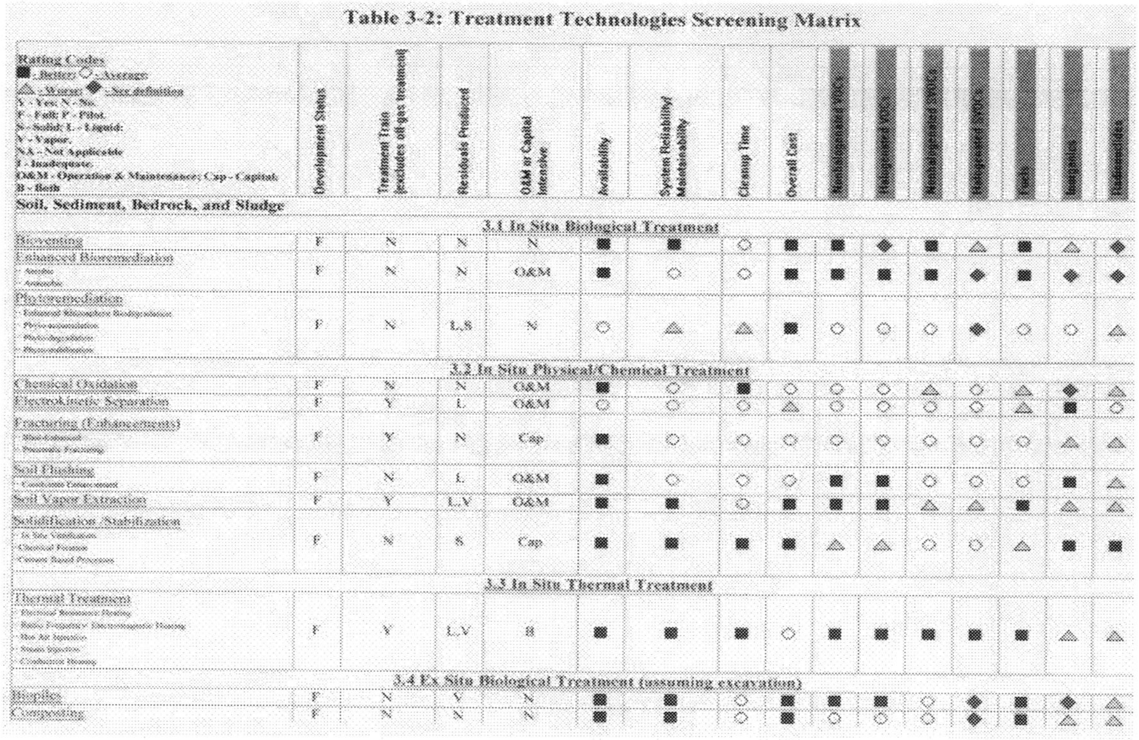

Table 3-2: Treatment Technologies Screening Matrix

Rating Codes
■ - Better; ○ - Average;
△ - Worse; ◆ - See definition
Y - Yes; N - No.
F - Full; P - Pilot.
S - Solid; L - Liquid;
V - Vapor.
NA - Not Applicable
I - Inadequate.
O&M - Operation & Maintenance; Cap - Capital;
B - Both

Soil, Sediment, Bedrock, and Sludge	Development Status	Treatment Train (excludes off-gas treatment)	Residuals Produced	O&M or Capital Intensive	Availability	System Reliability/ Maintainability	Cleanup Time	Overall Cost	Nonhalogenated VOCs	Halogenated VOCs	Nonhalogenated SVOCs	Halogenated SVOCs	Fuels	Inorganics	Radionuclides
3.1 In Situ Biological Treatment															
Bioventing	F	N	N	N	■	■	○	■	■	◆	■	△	■	△	◆
Enhanced Bioremediation	F	N	N	O&M	■	○	○	■	■	■	■	◆	■	◆	◆
Phytoremediation	F	N	L,S	N	○	△	△	■	○	○	○	◆	○	○	△
3.2 In Situ Physical/Chemical Treatment															
Chemical Oxidation	F	N	N	O&M	■	○	■	○	○	○	△	○	△	◆	△
Electrokinetic Separation	F	Y	L	O&M	○	○	○	△	○	○	○	○	△	■	○
Fracturing (Enhancements)	F	Y	N	Cap	■	○	○	○	○	○	○	○	○	△	△
Soil Flushing	F	N	L	O&M	■	○	○	○	■	■	○	○	○	■	△
Soil Vapor Extraction	F	Y	L,V	O&M	■	■	○	■	■	■	△	△	■	△	△
Solidification /Stabilization	F	N	S	Cap	■	■	■	■	△	△	○	○	△	■	■
3.3 In Situ Thermal Treatment															
Thermal Treatment	F	Y	L,V	B	■	■	■	○	■	■	■	■	■	△	△
3.4 Ex Situ Biological Treatment (assuming excavation)															
Biopiles	F	N	V	N	■	■	○	■	■	■	○	◆	■	◆	△
Composting	F	N	N	N	■	■	○	■	○	○	○	◆	■	△	△

Figure 2. Screen- EPA's Technology Assessment Matrix.

Table 2. Ease of Remediation & Contaminant properties Matrix.

Soil Type	Phyto Remediation	Bio Remediation	Air Stripping	Sparging	Excavation	Soil Washing	Thermal Treatment	Incineration	Stabilization	Encapsulation	In-Situ Thermal
Organic Constituents											
Gravel	g	g	g	g	g	g	u	u	g-f	u	u
Sand	g	g	g	g	g	g	u	u	g-f	u	u
Fine Sand	g	g	g	g	g	g	u	u	g-f	g	u
Silty Sand	g	g	g	g	g	g	u	u	g-f	g	u
Silt	g	g	g	f	g	g	g	g	g-f	g	g
Silty Clay	f	f	f	f	g	g-f	g	g	g-f	g	g
Clay	f	f	f-p	f-p	g	f	g	g	g	g	g
Fat Clays	f-p	f-p	p	f-p	g	f-p	g	g	g	g	g
Organic Muck	f	f-p	g-f	g-f	g-f	p	g	g	g	g	g
Perched Water Table	f-p	g-f	g	g	f	p	p	p	f	f	f
Lens Structure	f-p	g-f	g	g	g-f	p	f	f	f	f	f
Fractured Rock	p	g-f	g	g	f-p	p	p	p	f	p	p
Inorganic Consitutents											
Gravel	f	u	u	u	g	g	u	u	g-f	g	u
Sand	f	u	u	u	g	g	u	u	g-f	g	u
Fine Sand	f	u	u	u	g	g	u	u	g-f	g	u
Silty Sand	f	u	u	u	g	g	u	u	g-f	g	u
Silt	f	u	u	u	g	g	u	u	g-f	g	g
Silty Clay	f-p	u	u	u	g	g-f	u	u	g-f	g	g
Clay	f-p	u	u	u	g	f	u	u	g	g	g
Fat Clays	f-p	u	u	u	g	f-p	u	u	g	f	g
Organic Muck	f-p	u	u	u	g-f	f	u	u	g	f	g
Perched Water Table	f-p	u	u	u	f	p	u	u	f-p	f-p	f
Lens Structure	f-p	u	u	u	g-f	p	u	u	f-p	f	f
Fractured Rock	p	u	u	u	f-p	p	u	u	p	p	p

Key letters represent probability of successful outcome g= good, f= fair, p= poor, u = technology is unsuited for contaminant/ soil type mixture weren't and someone did not consider the costs and options involved in the disposal decisions, and the location of "Away", we are here considering the consequences of those decisions and trying to find out how to manage our wastes and how to remediate the disposal site which was the "Away".

1.1. REMEDIATING HIGH HAZARD CHEMICAL SITES WITH CONVENTIONAL TECHNOLOGY

Decontaminating a site where chemical agents have been applied is its own unique challenge. At a 1998 NATO conference in Bucharest, military decontamination was discussed and examined. Most of the military applications of decontamination were primarily for the equipment and materials, and it did not, at that time, consider any civilian reoccupation of the contaminated battle sites. I had the opportunity to learn some other interesting things about decontamination activities.

1.1.1. *Setting priorities*

When you are dealing with sites where agents have been applied, personnel safety and your is your highest priority. Special clothing, respiratory protection, sample collection and analyses procedures and decontamination procedures are all required before one begins characterization and any site work. The possibility of accidental exposure and accident prevention has to be continually considered.

1.1.2. *Reoccupancy?*

The second idea is that some sites may never be safely re-occupied. That has to be part of the planning process and decision-making. If the contaminants cannot be removed safely, and if the technology has a significant possibility of creating a primary or secondary exposure pathway, then leave the site alone! I'm not talking about capping a landfill or putting a few decimeters of cover over a site, I'm talking about posting permanent monuments and making the site a no-man's land. The precedent for this can be taken from the US Department of Energy Site Restoration program at Handford, Washington. At that site they have radioactive wastes which will not be safe to handle for tens of thousands of years. Because the potentially "safe" handling time is longer than written history, there was at one time a discussion of how to indicate the danger to future generations where the time frame is significantly longer than the written record of human history.

1.1.3. *Decontamination and remediation*

In a situation where military agents have been applied, there may be an equal likelihood that the military has also applied their own decontaminant materials. These do not necessarily render the agent neutral or environmentally safe, but merely less toxic. Indiscriminate use of the "decontamination" agents can be as environmentally damaging at the use of the agent.

We will consider two sites where agents have been applied. The first one is where VX has been used, and the second is where a radioactive compound has been used. The reason for the choices are because the compounds have substantially different properties, environmental effects, and remediation solutions.

VX is an organo-phosphate. The structure is shown below:

```
CH3      0                    CH(CH3)2
    \    ||                  /
       P - S - CH2CH2 - N
      /                      \
CH3CH2O                       CH(CH3)2
```

Now, when the decontaminant is applied, the resulting compound is a Methyl Phosphonic Acid or MPA and the principal mechanism is by hydrolysis. Depending upon whether one has a soil where decontaminants have been applied, EMPA (Ethyl Methyl Phosphonic Acid), and EA2192 {S-[2(diisopropylamioethyl)ethyl] methylphosphonothioic acid} may also be present. EA2192 is almost as toxic as VX, and EMPA or whether the hydrolysis is natural, EMPA, which is almost as toxic as VX may be present as well. The mechanisms for degradation and the pathways have been documented by Yang and others. If military decontaminants have been applied, the soil may also have a very high pH and Chloride and Calcium content because Calcium Hydroxide is one of the major components of VX decontamination products.

While the literature indicates that the half life of various degradation products of VX have a soil half life of greater than 100 days, except MPA, , the relative toxicity of the compounds suggests that the contact of personnel and contaminated materials should be prevented, and the disposition of the site must be final and address surfaces which are difficult to decontaminate because of size or porosity or both. The material is too dangerous to permit any chance of an accidental release from a partially decontaminated site.

1.2. THE THERMAL OPTION

The apparent solution to this is thermal decontamination, incineration, thermal desorption, or oxidation. However, when you are done you will have oxidized, or volatilized all the contaminants in the soils, rocks, etc, and you will have effectively sterilized the soils. Depending upon the temperature, most of the carbon and organic matter in the soils will be gone – converted to CO2, and some inorganics will be in highly oxidized sates or as oxide salts. Incineration of various types of organic chemicals has led to fears about Dioxin generation, but this is generally controlled by secondary flame temperature and residence time. In some cases, thermal oxidation has converted Chromium to a +6, and in other instances, Arsenic has been volatilized. After thermal treatment, the soil will have to be reactivated if it is to be productive. This generally requires fertilization and the addition of nutrients, and organic matter.

For a high toxicity site, a rotary kiln with a very good secondary flame chamber, and an excellent scrubber and baghouse system should be employed. A typical system is shown below in Figures 3 & 4. The system shown is capable of destruction of PCB's and other highly toxic contaminated soils at a rate of about 20 tons per hour. The primary does not need to get overly hot, say about 450oC or so, but hot enough to vaporize the compounds from the soils and make sure that they are released. The real work of toxicity destruction is performed in the secondary destruction chamber, where one insures that the compounds are totally destroyed by the combination of Time, Temperature and Turbulence. After that, a fast quench helps prevent the formation of secondary dioxins, and the baghouse and the scrubber do the rest.

The wastes generated include scrubber water which can contain some particulate and calcium carbonate or other chemical sludges, and various types of salts, and some dusts from the baghouse and cyclone which are recombined with the soil and are tested for various contaminants.

The next few pages show pictures of the type of equipment to be used.

For a large area, large equipment is required. The equipment needed to manage a 20 tons per hour (17.8 MT/ Hour) comes in 16 truck trailers, and requires about five weeks to set up, if all the permits and utilities and transportation equipment are in place. The logistics behind operation of a large scale system are no small matter either. The burner must be kept hot, and it uses large quantities of propane which is generally brought in by truck. The electrical power is run by diesel generator, also brought in by truck. Then there is the fuel for the materials handling equipment – front end loaders, trucks, graders, etc. Let us not forget the chemical supplies for the laboratory and the laboratory required to provide the analyses of the wastes and the raw materials.

Add to this the fact that the soil must be brought to the site, and you begin to see what a logistical challenge this type of operation can be. For example, if the soil has an average density of about 1.4, that means that a cubic meter will weigh about 1.4 MT, and a 20 MT load on a truck can be about 12 CuM, if you have a truck(s) of that capacity. At maximum capacity one would have to have between 1 and 2 trucks per hour around the clock for the duration of the burn.

On a site, materials handling and dust control are always the issues which must be faced. Because the site contains MPA which can pose both an environmental and a direct health risk, the entire excavated site will have to be wet down regularly for dust control, and despite the mud problems this creates, water and runoff from the site must be collected, tested, treated, prior to its release. The same thing must happen for the treated soils.

1.2.1. *Worker Stress is a factor*

Equipment operators and technicians may be working in Level A (SCBA Chemical Suits), and workers must wear several layers of protective gloves and clothing to prevent accidental contact with contaminated material. Heat stress & exhaustion will be a problem for workers. Before maintenance is conducted, all equipment must be thoroughly decontaminated, and everything, including clothing and safety equipment will have to be decontaminated as well. The same is true for the trucks and drivers. If your site is more than a few hundred meters on a side, you must face the task of distance hauling. That means special packaging or containers to prevent the soils from being blown off the truck. What do you do if a truck breaks down or has a flat tire or needs routine maintenance. Special decontamination is required before the mechanic can even begin to approach the truck.

Depending upon the size of the site, it may be easier to take the incinerator to the site than it is to transport the soils to the incinerator. That, however, is a decision which is often based upon operating constraints, equipment, and cost. Most often, the size and portability of the incinerator is the governing factor.

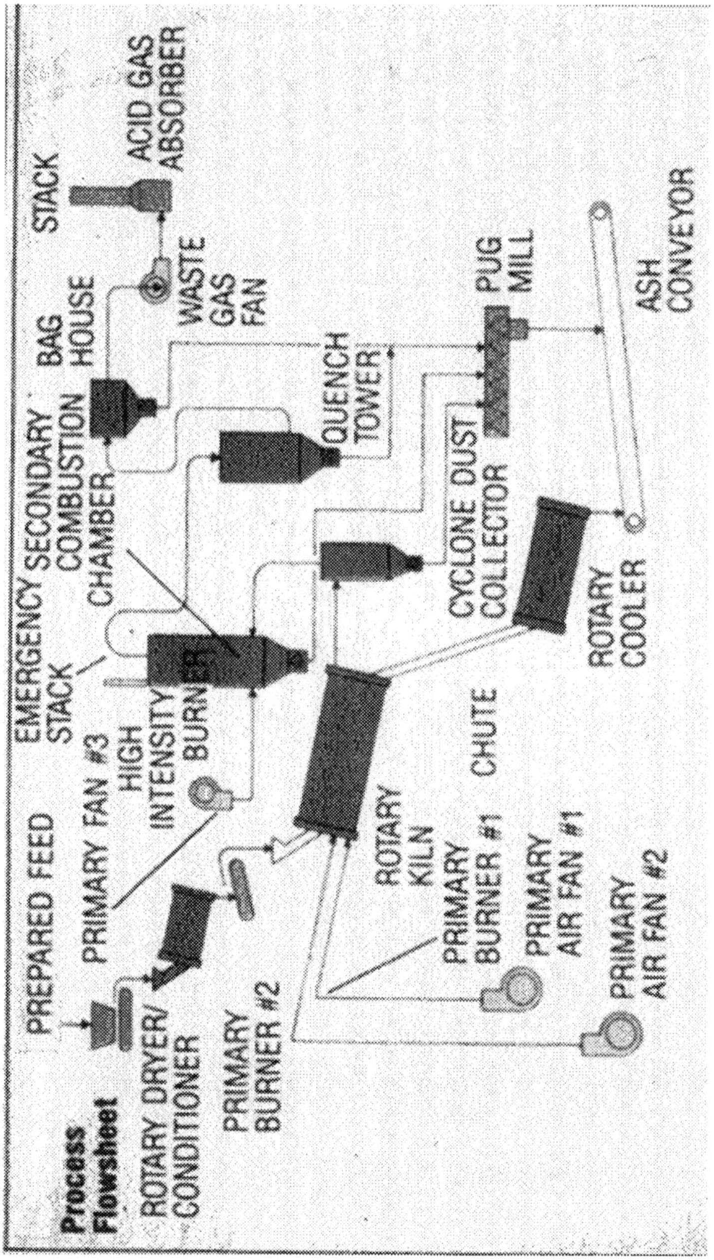

Figure 3. High Efficiency Thermal Destruction Unit.

Drawing Courtesy: <u>T. F. McGowan Associates, Atlanta, GA</u>

Figure 4.
Photograph Courtesy Compass Environmental, Inc., Atlanta, GA

In addition to this, there is the level of worker stress. While that is psychological, that stress is real, and there is an awareness of the constant danger, and the potential for accidents. There is a continual low level awareness of what the worker is exposed to and the potential for disaster, even in the "safe" areas. Extended worker vacation is often necessary if only to relieve psychological stress.

2.4. ALTERNATIVES TO INCINERATION: SOIL WASHING

A few years ago I was asked to look at a proposal for soil washing in Belarus to remove the contamination created by Chernobyl. The proposal was fatally flawed for a number of reasons primarily dealing with logistics and materials handling. The proposal was developed by a German firm working in conjunction with Rockwell International, a large US Government contractor. What they were proposing was to take about 15-30 cm of soil from several thousand hectares of Belarus which had been contaminated with particulate Uranium, Strontium, and other radioactive materials,

The proposal was fatally flawed by failing to address some of the most critical elements of soil washing. Soil washing is essentially a hydraulic flotation process which removes the fines from the soil. Depending upon the soil, that can account for between 5% and 15% of the volume processed. The process is strictly one of density settling and stokes law is followed in the separation process. What you wind up with is a clean and sterile soil because the organic materials in the soil have a density of between 1.2 and about 2.0 and the clays, and some of the silts, because of their particle size are removed from the soil. In the case of Belarus soils, this also removed about 60%-80% of the fine radioactive materials, but that was not the problem. The problem was one of scale and residuals.

2.4.1. *Belarus considerations*

Consider a site for a moment which has an area of 10 square kilometers. That's 1000 hectares, or 10,000,000 square meters. Now take 15- 30 cm from the soils. You will process a pile of soil of between 1.5 and 3 million cubic meters, excluding the buildings, trees, and other things you might have trouble dealing with. Your first challenge is the transportation. This will require a minimum of 105,000 – 210,000 truck loads, just to move the soils to a processing site. At 2 round trips per hour, that's about 40 tons per hour per truck, and assuming a fleet of 1000 trucks, the process would require between 50 and 100 days, if you can move 1000 trucks in an orderly fashion and that you had processing equipment sufficient to process 40,000 tons of soil per hour. Most processing units have an upper limit of 50 tons per hour, so you would require about 800 processing units. We are only discussing 10 square kilometers here, not the entire contaminated area.

Transportation and logistical elements aside there is still the waste problem. Soil washing will generate between 5% and 15% of the treated soil as wet solids. That means that with our 1.5-3.0 million cubic meters, we will have a minimum of about 25,000 CuM of soil, to a maximum 450,000 Cu.M of soils which contain the radioactivity. Where you put this pile of waste materials was never considered in the proposal.

Because the World Bank was being asked to underwrite the financing for the soil cleanup, we investigated the costs: A single hectare of soil washing was about $230,000 and that was just the processing, and not the transportation, storage or waste disposal. At that time, the value of the cash crops in Belarus was less than $100 per hectare. The fact that the project would have required over 100 years to complete was also conveniently ignored in the proposal, as was the roadway and bridge and other infrastructure improvements required for the project. There just was not enough revenue in all of Belarus to support the cleanup, despite the Government's proposed 17% tax increase on all industrial activity. (In fact, had the tax ever been put in place, it might have proved a disincentive to further economic and industrial development.)

The project was dropped. Even if phytoremediation tried on this site and if it were considered successful in remediating the soils, the fundamental question of what to do with the contaminated plants and trees would still pose a huge challenge.

3. Cost elements in cleanup

The cost of cleanup is always a factor when you consider the potential uses of the land. The cost of the cleanup is why the US EPA has moved to the Brownfields program, and to "natural attenuation", although you probably could not get anyone in the EPA to admit it. Brownfields is a program to rejuvenate selected sites for industrial, commercial and residential uses by providing a barrier between the contamination and the new occupants. Natural attenuation is just a way of letting natural forces of dispersion, and natural biological action degrade the contaminants. Cynically, another name for natural attenuation is "paperwork remediation" in an effort to generate sufficient studies to prove that the contaminants can be left alone safely.

3.1. INCINERATION COSTS

A good cost estimate to site incineration is about $28/ MT for large sites of several hundred to several thousand tons. That includes limited "extra" safety precautions and limited waste handling, perhaps up to half a kilometer from the site.

3.2. GROUNDWATER COSTS

The groundwater must also be treated to remove contamination. The costs of this may run as high as $160 per cubic meter. This is based on a figure of $.16 per liter for highly contaminated water. The estimate includes equipment costs and amortization. Equipment needed may include filtration, centrifuges, sedimentation, chemical treatment and any materials disposal.

If the deep groundwater on the site is contaminated, even for a small site, the costs may exceed $55,000 to $300,000. The more complex the site is, the greater the costs. And these are extremely rough figures.

3.3. RIVER and SWAMPS

If river cleanup is involved, including dredging, the first estimate of the costs likely to be incurred is on the order of $6.6 Million per kilometer. This includes dredging the bottoms, cleaning and excavating the sides, and thermally treating the sludges after they are dewatered. The sediments will be pumped out at about 2% solids and have to be put into a contained area, the liquid treated, and the solids dried and excavated.

If a wetland is involved in contamination, the first question is can it be restored? If it can you may be looking at a figure of twice that for river restoration.

3.4. LABORATORY AND ANALYTICAL COSTS

Finally you have your laboratory costs. For even a modest site involving petroleum where you are running 30-50 samples per month, the cost can exceed $300,000 per month. A good and fully equipped laboratory can exceed $500,000 in costs, and a single sample of soil by GC/MS can exceed $2500 when you include labor to collect the sample, transportation, chain of custody, and costs for analyses. If you have a radioactive program, the costs can double or triple.

The analytical protocols and the handling and data security protocols are extremely important. I recently was reminded that if you have to ship samples to another country for analysis, that may require additional time, additional expense, and the customs, agriculture, and drug enforcement officials may want to thoroughly inspect your shipment thoroughly. Your laboratory may also require a special license to handle soil and water samples from out of the country. The logistics cannot be ignored, except at your peril.

4. Summary

Site cleanup is very expensive. One of the reasons why phytoremediation is of extreme interest is the cost factor and the apparent benefits. If phytoremediation cannot be accomplished on certain sites, the alternative solutions can be quite expensive. The development of phytoremediation technology is being encouraged because of the economic and environmental benefits.

FROM LABORATORY EXPERIMENTS TO LARGE SCALE APPLICATION – AN EXAMPLE OF THE PHYTOREMEDIATION OF RADIONUCLIDES

PETR SOUDEK, ŠÁRKA VALENOVÁ, DAGMAR BENEŠOVÁ AND TOMÁŠ VANĚK
Department of Plant Tissue Cultures, Institute of Organic Chemistry and Biochemistry, Czech Academy of Science, Flemingovo nám. 2, 166 10 – Prague 6, Czech Republic

*To whom correspondence should be addressed.

Abstract- Phytoremediation is widely viewed as an environmentally sound alternative to the destructive physical remediation methods currently practised. Plants have many endogenous genetic, biochemical, and physiological properties which make them ideal agents for soil and water remediation. Significant progress has been made in recent years in developing native or genetically modified plants for the remediation of contaminated environments. Because elements are chemically stable, phytoremediation strategies for radionuclide and heavy metal pollutants focus on above-ground hyper-accumulation. Soil contaminated with radionuclides pose a long-term radiation hazard to human health through exposure mainly via the food chain. Remediation of radionuclide-contaminated soils has become increasingly important. Removal of the contaminated surface soil (often down to 40cm) or immobilization of radionuclides in soil by applying mineral and chemical amendments are physically difficult and not cost-effective. Reducing plant uptake of radionuclides, especially ^{137}Cs and ^{90}Sr by the provision of competitive cations in chemical fertilizers is advantageous in large scale, low-level contamination incidents on arable land, and has been widely practised in Central and Western Europe following the Chernobyl accident. In this contribution we summarize results obtained from laboratory, greenhouse as well as field experiments focussed on phytoremediation methodologies for the removal of radionuclides from both the soil and the water.

N. Marmiroli et al. (eds.), Advanced Science and Technology for Biological Decontamination of Sites Affected by Chemical and Radiological Nuclear Agents, 139–158.

Keywords: phytoremediation, radionuclides, 137-caesium, 90-strontium, 125-iodine, uranium, radium, uranium mill tailings, biomonitoring

1. Introduction

Radionuclides are high risk pollutants of significant impact to the environment, food chain and human health. The removal of radionuclides from the environment is important for human health because the ionizing radiation they emit is harmful. Large irradiation doses can cause extensive cellular damage and result in cell death, but even low levels of chronic exposure are known to be responsible for genetic changes, pre-cancerous lesions, benign tumours, cataracts, skin changes and congenital defects.

Uranium ore processing factories, nuclear power plants, nuclear bomb testing areas and accidents in nuclear facilities have introduced large quantities of radionuclide into the environment. The Chernobyl accident in 1986 has resulted in long-term environmental radiological contamination throughout the whole of Europe, and the release of ^{134}Cs and ^{137}Cs (respectively, 54 and 85PBq) are the most important of these contaminants. The most highly contaminated areas are in the three former Soviet Union republics of Belarus, Ukraine and the Russian Federation, where an area of about 150,000km^2 is contaminated at more than 1Ci/km^2 (37kBq/m^2). Substantial areas of other European countries have been similarly affected, e.g., Sweden (12,000km^2), Finland (11,500km^2), Austria (8,600km^2), Norway (5,200km^2) and Bulgaria (4,800km^2), while lower activity (<1.Ci/km^2) has been distributed widely over central and western Europe (EU publication, 1998; Bennett et al., 2000). Various studies have analysed the dynamics of ^{137}Cs in natural and semi-natural environments (Nisbet, 1993; Sadolko et al., 1995; Arapis et al., 1997; Bunzl et al., 1997; Velasco et al., 1997; Belli, 1998; White et al., 2002).

The use of plants to remediate soils, sediments, surface and ground waters contaminated by radionuclides and/or toxic elements has been widely reported, especially for soils (Nishita et al., 1958; Broadley et al., 1999; Dushenkov et al., 1999; Lasat et al., 1998; Watt et al., 2002; Whiley et al., 2001; Soudek et al., 2004a) and waters with low levels of contaminant (Cornish et al., 1995; Soudek et al., 2004b; Vaněk et al., 2001; IAEA, 1989; Soudek et al., 2006b). Equivalent methodology, based on plant uptake of radionuclides (radiophytoremediation) or toxic element contaminants from water, has occasionally been used for biomonitoring (Wolterbeek van der Meer, 1996). Radionuclide decontamination does not involve a toxicity of high concentration as is the case for toxic metals, but rather the problem is the radioactivity itself. The mass of radionuclide contaminant is very low and usually falls below the detection limit of standard analytical methods (see Table 1).

Phytoremediation methods for radionuclide decontamination do not involve hyper-accumulators, except possibly for uranium. Plants require a long period of contact with a contaminant to evolve the ability to hyper-accumulate, and most uranium ores are located underground and so are not in contact with plants. Soils with high concentrations of uranium are present only where uranium is or has been mined or processed, but these have only been in existence for a few decades.

Table 1. Radionuclide specifications. (Amount of radionuclide is calculated on activity 10 times higher than average natural background which is 0.02 Bq/g DW) (DOEEM, 1998).

Radionuclide	Specific activity [Ci/g]	Amount of radionuclide
^{60}Co	$1.131 \times 10^{+03}$	11.9 fg
^{90}Sr	$1.364 \times 10^{+02}$	99.1 fg
^{125}I	$1.737 \times 10^{+04}$	0.8 fg
^{137}Cs	$8.698 \times 10^{+01}$	0.2 pg
^{226}Ra	9.887×10^{-01}	13.7 pg
^{238}U	3.362×10^{-07}	40.2 µg

2. Uptake and translocation of radionuclides under hydroponic and *in vitro* conditions

We have studied the uptake, translocation and distribution of ^{137}Cs, ^{90}Sr and ^{125}I by sunflower, poplar and reed. Attention has been focused not only on the time course of uptake from a radioactive hydroponic solution, but also on the distribution of radioactivity across plant tissues. Sunflower has also been used to identify the influence of K^+, Ca^{2+} and NH_4^+ on ^{137}Cs and ^{90}Sr uptake and accumulation, with the aim of evaluating the effect of these ions which are normally present in the soil.

None of the three plant species compared (among which poplar was the best-performing) are generally considered as hyper-accumulators, since they absorb up to 1% of their dry weight as metals (Baker & Brooks, 1989). However, an ability to hyper-accumulate is not an absolute prerequisite for the effective uptake of radionuclides, because these are present in nature only in small mass concentrations. The most critical plant property for remediation of radionuclide contamination is a high growth rate and the ability to generate large amounts of biomass in a given environment (Fig. 1).

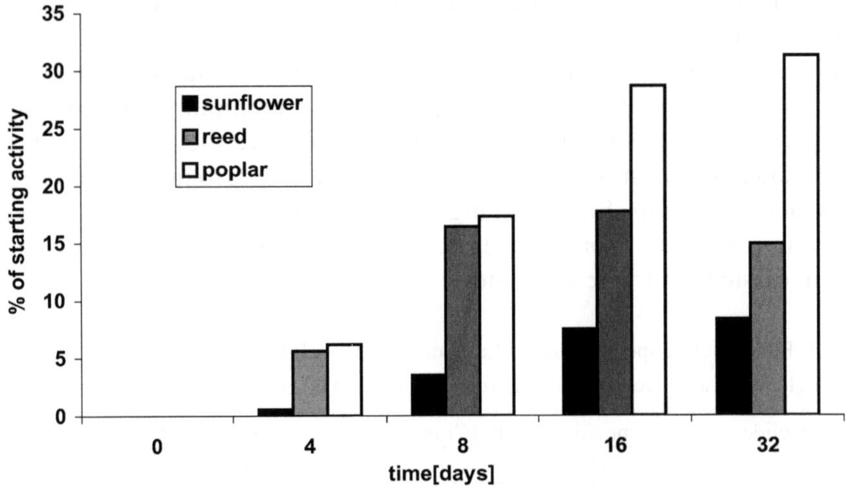

Figure 1. Accumulation of ^{137}Cs by different plant species. (Soudek et al., 2004b).

Because of their similar chemical properties, strontium and calcium are taken up in a similar way by plants (Mengel and Kirkby, 1978; Soudek et al., 2006d). In our experiments with sunflower, we have used calcium concentrations present in agricultural fertilizer. The highest uptake of ^{90}Sr was observed at 8mM Ca^{2+} (Fig. 2). Higher concentrations of calcium ions resulted in a lower accumulation of the strontium nuclide. Tyson *et al.* (1999), Shaw (1993) and Roca and Vallejo (1995) all concluded that the less calcium present in the soil, the more strontium is accumulated, but our results are not in full agreement with this conclusion. The high accumulation of ^{90}Sr in the presence of low concentrations of calcium ions could be due to its very low concentration, whereas calcium was present at higher concentrations.

3. Autoradiography

Autoradiography allows the localization of radionuclide within plant tissue, which is important for an understanding of the uptake and translocation mechanisms. It also represents a selection method for the efficiency

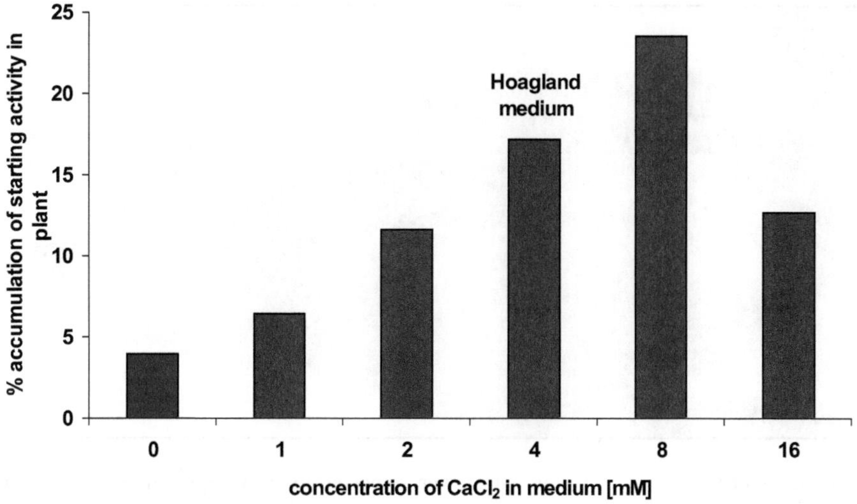

Figure 2. Accumulation of ^{90}Sr by sunflower cultivated on the hydroponic medium with different concentration of calcium ions. (Soudek et al., 2006d).

of both phytoextraction (maximum accumulation in the aerial parts) and phytostabilization (maximum accumulation in the below-ground parts, which effectively minimises contamination of the food-chain).

The distribution of ^{137}Cs in sunflower appears to be strongly dependent on its transport together with water and nutrients. The strongest activity is concentrated in young leaves, leaf veins, nodal segments, in the youngest parts of the plant, and in the internodal segments. These are sites either with the highest metabolic activity, where cell growth and differentiation take place, or those, like the leaf veins (Fig. 3), which actively participate in the transport of substances. In contrast, the roots show little activity, even though they are involved in the transport of water, nutrients and the nuclide. Cs is therefore highly mobile into the upper parts of the plant, and is selectively retained in the above-mentioned structures. Almost no Cs was present in the root tip, although this represents one of the most active sinks in the plant. The stem between the roots and the nodal segment with the cotyledons also shows a higher level of activity (Soudek et al., 2004b).

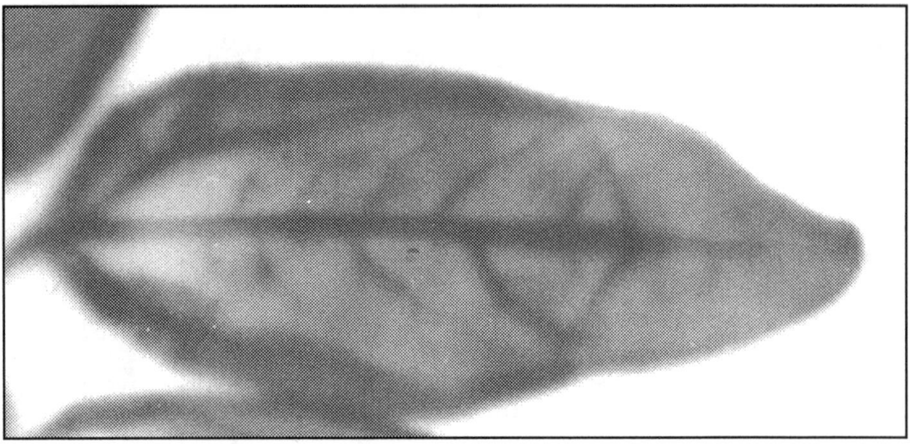

Figure 3. Distribution of ^{137}Cs in sunflower exposed to 5 $MBq.l^{-1}$ after 32 days of cultivation (detail of leaf). (Soudek et al., 2006d).

^{137}Cs was accumulated to some extent in all plants and their tissues studied, mainly at the sites of cell growth and differentiation, where metabolism is high and the demand for water and nutrients significant. This particularly applies to nodal segments, leaf tips, young leaves and young shoots. Cline and Hungate (1960), Smolders and Shaw (1995) and Zhu et al. (1999) all reported higher Cs concentrations in roots than in leaves and stems, but our results show an approximately 1:1 ratio of Cs concentration in roots and aerial parts. With the exception of some sites with extremely high Cs concentration, the nuclide was equally distributed over all plant tissues, in agreement with Menzel and Heald (1955). Buyss et al. (1995) measured the circulation of ^{137}Cs in spinach and found that 75%–95% of the nuclide was recycled into the roots. The mobility of Cs in the plant was apparently great (Soudek et al., 2006d).

The ^{90}Sr nuclide was relatively equally distributed throughout the plant, although a higher level of activity was present in the leaf veins, stem and central portion of the root. Small sites of activity scattered over the leaves probably represent the position of stomata (Fig. 4). Sr has a similar function as calcium in plants, and is important for the action of stomata, where its concentration in the guard cell cytosol is influenced by ABA (Procházka et al., 1998). Sr is largely transported from the roots into the aerial parts of the plant, as documented by Entry and Emmingham (1995) in *Eucalyptus tereticornis* seedlings, and by von Fircks *et al.* (2002), who showed that the ^{90}Sr concentration in *Salix viminalis* was three times higher in leaves than in roots, stems or cuttings. Similarly, Tyson *et al.* (1999) showed that 80% of all the ^{85}Sr provided to the roots of *Pteridium aquilinum* was transported

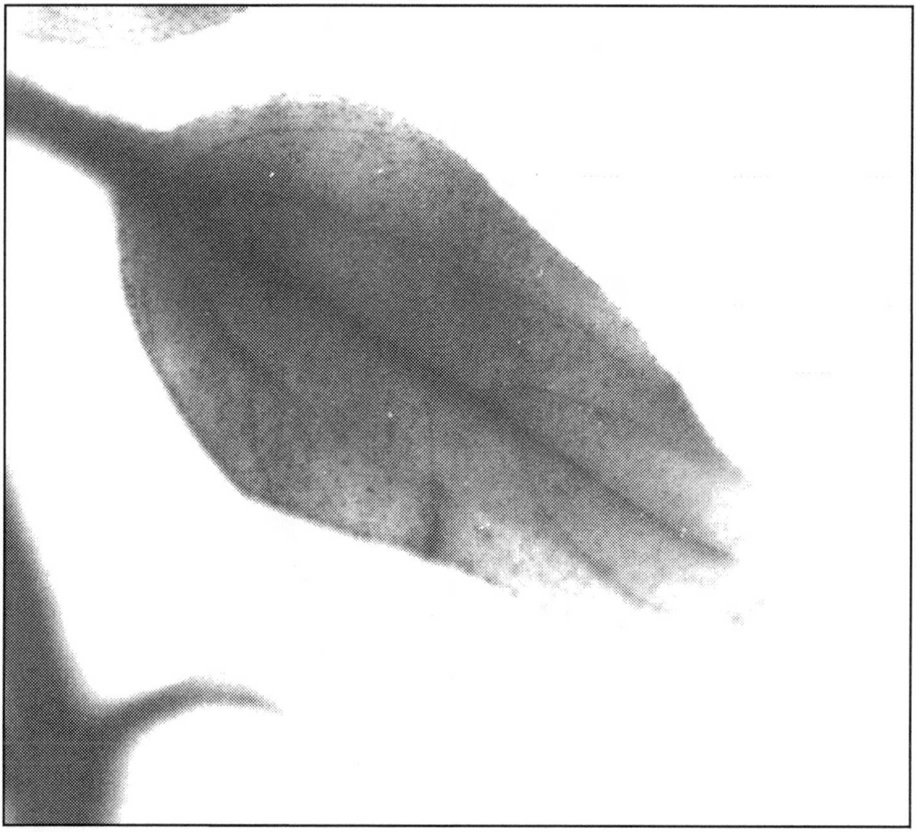

Figure 4. Distribution of ^{90}Sr in sunflower exposed to 5 $MBq.l^{-1}$ after 32 days of cultivation (detail of leaf). (Soudek et al., 2006d).

hydroponically cultivated sunflower showed a different pattern of distribution from the above-mentioned plants which were grown on soil contaminated with Sr nuclide (Soudek et al., 2006d).

The shoots (and especially the internodal segment between roots and cotyledons) and the stalk of the upper part of the plant accumulated most of the carrier ^{125}I (Fig. 5), but the leaf veins showed a slightly elevated level. It appears that ^{125}I accumulates in tissues associated with high metabolic activity, in organs with a transport function, and in tissues carrying a high concentration of starch (cotyledons and the internodal segment between the cotyledons and roots). Although roots represent the site of I uptake and root tips exhibit a high metabolic activity, no significant ^{125}I accumulation was detected in the root system. An elevated accumulation of activity in the presence of non-carrier ^{125} I was found in the roots, and a smaller level in

Figure 5. Autoradiography of sunflower cultivated with ^{125}I ([A] hydroponic medium without 0.1 mM KI, [B] hydroponic medium supplemented by 0.1 mM KI). (Soudek et al., 2006b).

nodal segments. The increasing activity in the root systems can be explained by the presence of a low concentration of I, and the possibility of I binding in root cells (Soudek et al., 2006b). These results are in agreement with those of Muramatsu et al. (1995).

Radionuclides can be also used to study the accumulation and degradation of organic pollutants. In our experiments we have followed the uptake and degradation of ^{14}C labelled TNT by wetland plants (Nepovím et al., 2005), and showed that about 63% of the ^{14}C localized in the roots of *Ph. australis* was bound (Fig. 6) and the remainder was acetone-extractable. An HPLC analysis of the acetone extract failed to detect any TNT, showing that all TNT had been chemically transformed. Thus TNT was not adsorbed on the root surface. In similar experiments performed in wheat (*Triticum aestivum*), Sens et al. (1999) found that 57% of the ^{14}C taken up was bound

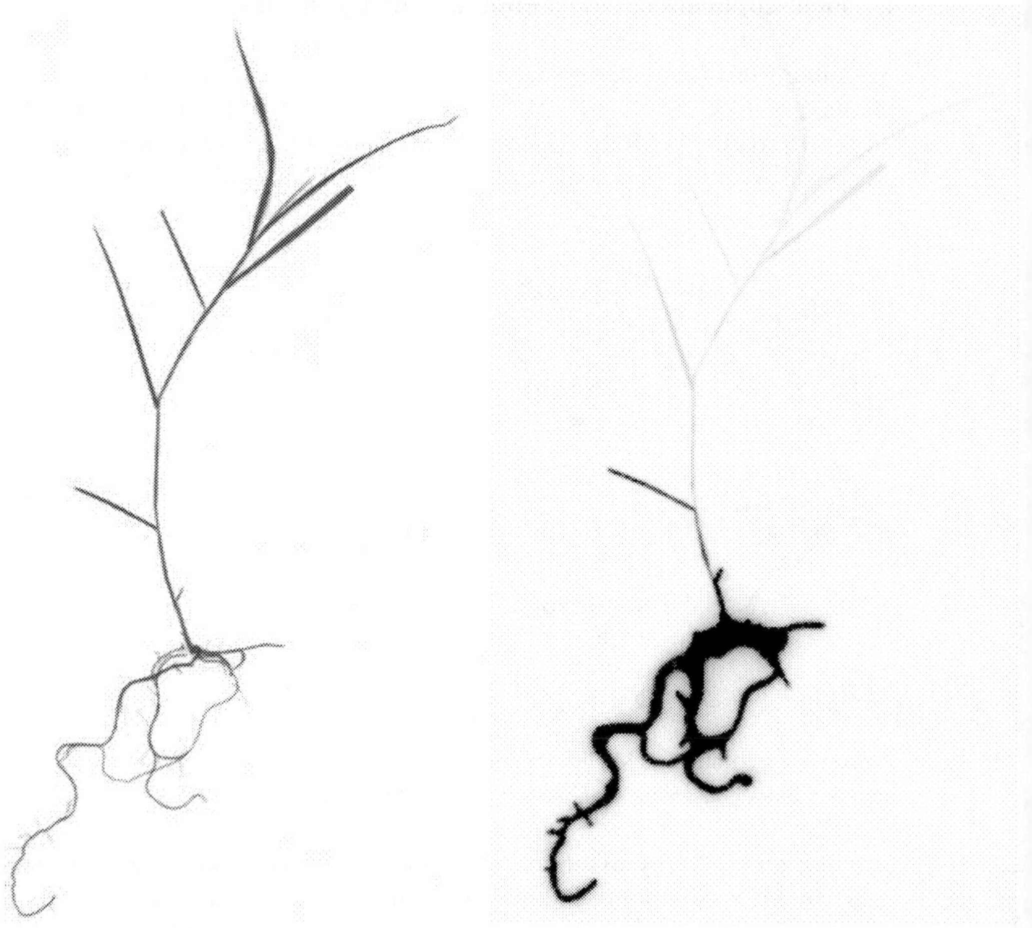

Figure 6. Distribution of radioactivity in reed. A photograph of plant (a) and its autoradiogram (b). Plants were treated by [^{14}C]TNT (1.02 MBq/L) for 14 days. A plant was washed, dried and exposed to film for 3 weeks. (Nepovím et al., 2005).

to the cell wall and the rest was detected in the cytoplasm in the form of soluble compounds. A lignin fraction contained 27% of the ^{14}C taken up, a proportion comparable to that achieved in bean (*Phaseolus vulgaris*) (Sens et al., 1998). These studies show that there is no general pattern for the metabolism of TNT in plants, especially considering the wide spectrum of degradation products that can be formed. A major degradation pathway can be proposed, but the formation of end products differs between plant species, depending on which enzymes involved in the detoxification processes are available.

4. Field application – a uranium mill in Mydlovary

A uranium ore reprocessing factory in South Bohemia near the village of Mydlovary (49° 5' N, 14° 21' E) was in operation from 1962 to 1991 (Fig. 7). About 17Mt of U-ore of low uranium content (0.18%) were processed over this period, using two technologies: about 76% by acid

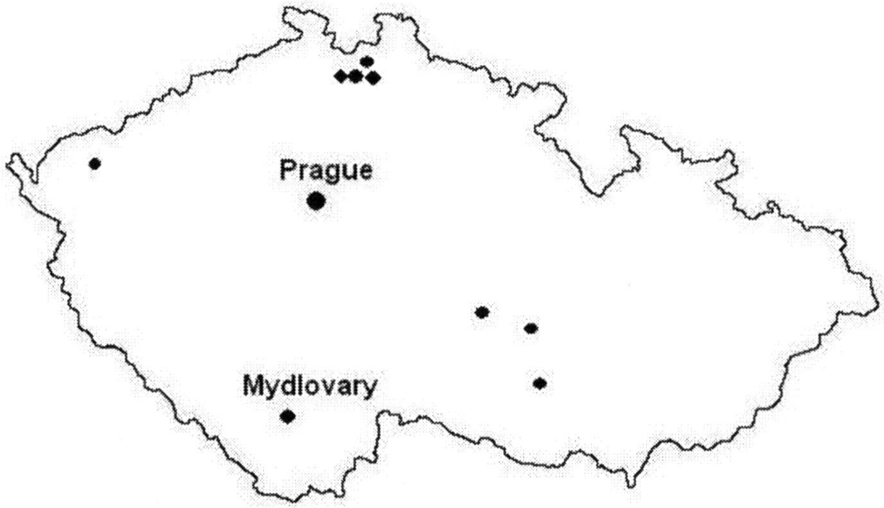

Figure 7. Localization of Mydlovary contaminated site on the map of the Czech Republic (the other black dots are the uranium mines). (1:4 000 000).

leaching and 24% by alkaline extraction. The sludge beds are distributed over an area of about 2.3km^2. The mill tailings from both types of extraction were deposited on waste dump K1, a botanical survey was performed and the ^{226}Ra activity content determined in the predominant plant species present, before the dump was covered with sterile soil in the summer of 2004. The contaminated mill tailings contained particles ranging in size from 45 - 75μm, small amounts of SiO_2, but mostly was composed of very complex clay-like silicates of Na, Ca, Mn, Mg, Al, and Fe and metallic oxides. The average activity is about 23Bq ^{226}Ra per g DW and the pH is 8.06 (details in Fig. 8).

The aim of plant collection on the K1 waste dump was to identify naturally widespread plant species able to grow on contaminated areas and/or accumulate large amounts of radionuclides, and which would be suitable candidates for radiophytoremediation purposes. In all, 44 plant

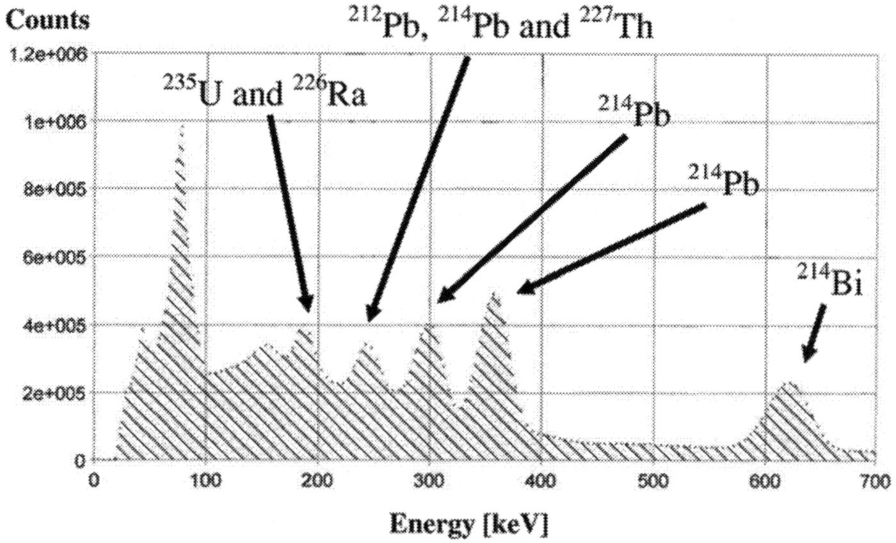

Figure 8. Measured a soil sample spectrum in decay equilibrium.

species were tested, including both herbaceous and woody plants, featuring three major plant genera (*Asteraceae, Fabaceae* and *Rosaceae*). Excluders-specifically *Trifolium repens, Amanita phalloides, Persicaria amphibia* and *Tripleurospermum inodorum*-can grow in the presence of high activity, but take up very little ^{226}Ra in their biomass. Accumulators, such as *Potentilla reptans* (4.09Bq/g of DW, *Mentha arvensis* (4.00Bq/g) and *Daucus carota* (3.70Bq/g), took up significant amounts of ^{226}Ra, but none of these species can be characterized as radiohyperaccumulators, as the highest amount of Ra accumulated was only about 110ng per kg DW, which, according to Baker and Brooks (1989), is far below the hyperaccumulator limit. A different criterion for the separation of the species into two groups focusses on their utility for medicinal purposes. About a half of these species have been classed as medicinal plants (according Bruneton, 1995); while some accumulate ^{226}Ra (e.g. *Mentha arvensis, Hypericum perforatum, Fragaria vesca, Centaurium erythraea*), other are excluders (e.g. *Urtica dioica, Verbascum thapsus, Potentilla anserina, Tanacetum vulgare*). Almost all the *Asteraceae* and *Fabaceae* species accumulate ^{226}Ra at around the detection limit, while the *Rosaceae* species accumulate about 2Bq/g DW (Soudek et al., 2004a; Soudek et al., 2006a).

5. Greenhouse experiments

A series of greenhouse experiments was performed in March to July 2004. Early results showed differences between crops species in their efficiency of radionuclide uptake. The activity of the soil was about 13Bq ^{226}Ra/g DW, and the highest accumulators were *Amaranthus tricolor* cultivar Early Splendor and *Lupinus polyphyllus* (respectively, 2.16 and 2.20Bq ^{226}Ra/g DW). No significant differences between cultivars of a single plant species were found, but there were differences in the accumulation of ^{226}Ra between species within a genus (*Lupinus* sp.) (Table 2). Two *Lupinus* sp. (*albus* and *luteolus*) were used to test the soil/plant transfer of ^{226}Ra

Table 2. The activity of plants cultivated in greenhouse on soil mixture (K1) with activity 13 Bq ^{226}Ra/g DW. (Soudek et al., 2004a).

Plant species	**Activity ± S.D. [Bq ^{226}Ra/g DW]**
Linum usitatissimum „Atalante“	0.45 ± 0.013
Linum usitatissimum „Jitka“	0.35 ± 0.018
Cannabis sativa “Beniko“	0.42 ± 0.008
Cannabis sativa „Juso-11“	0.40 ± 0.016
Cannabis sativa „Silesia“	0.52 ± 0.010
Amaranthus hypochondriacus „Pygmy Torch“	0.74 ± 0.020
Amaranthus tricolor „Early Splendor“	2.16 ± 0.071
Amaranthus tricolor	0.59 ± 0.022
Amaranthus caudatus „Atropurpureus“	0.76 ± 0.020
Phaseolus vulgare „Bobis Nano“	0.95 ± 0.039
Phaseolus vulgare „Aida Gold“	0.55 ± 0.031
Pisum sativum „Ambrosia“	0.32 ± 0.030
Pisum sativum „Gloriosa“	0.43 ± 0.019
Capsicum annuum „Berta“	0.40 ± 0.026
Capsicum annuum „Drákula“	0.46 ± 0.025
Capsicum annuum „Maryša“	0.56 ± 0.021
Lycopersicon lycopersicum „Albertovské“	0.86 ± 0.027
Lycopersicon lycopersicum „Stupické“	0.57 ± 0.023
Lycopersicon lycopersicum „Start F1“	0.32 ± 0.021
Lupinus albus	0.72 ± 0.039
Lupinus luteolus	1.24 ± 0.093
Lupinus polyphyllus	2.20 ± 0.097
Daucus carota “Karotina“	1.10 ± 0.016
Sinapis alba „Zlata“	0.55 ± 0.012
Helianthus annuus „Obrovská“	0.41 ± 0.016
Brassica oleracea „Opera“	0.75 ± 0.030

(according PhytoDec, 2004). Polluted soil was mixed with clean soil to obtain an activity range of ^{226}Ra corresponding to the phytoextraction starting point and remediation target. Then, medium sized pots were filled with the equilibrated soil and the activity of ^{226}Ra was determined. The target plant species were sown into the pots, and after two months of growth, ^{226}Ra content in the plant tissue was measured. The Log-log linear relationship between plant ^{226}Ra activity and the potentially available soil ^{226}Ra activity can be represented by:

$$\log[^{226}\mathrm{Ra}]_{\textit{Lupinus albus}} = -4.8736 + 1.155 * \log[^{226}\mathrm{Ra}]_{\mathrm{soil}}$$
$$R^2 = 0.9065$$

$$\log[^{226}\mathrm{Ra}]_{\textit{Lupinus luteolus}} = -0.1351 + 0.6835 * \log[^{226}\mathrm{Ra}]_{\mathrm{soil}}$$
$$R^2 = 0.7546$$

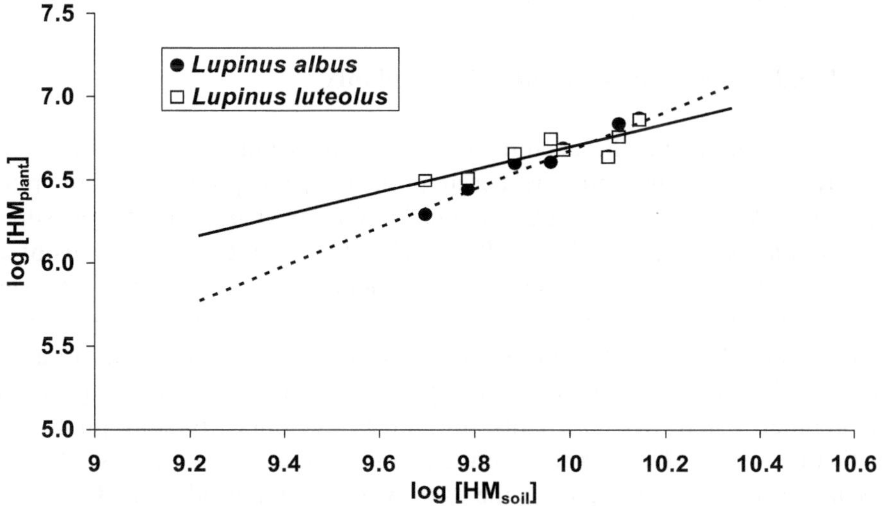

Figure 9. Greenhouse experiment - soil / plant transfer of ^{226}Ra.

Following this, the duration for phytoremediation was calculated. For this purpose, we took the activity of the soil in Mydlovary to be 21.75Bq ^{226}Ra/g DW, the amount of polluted soil at Mydlovary to be 1000t, and the annual *Lupinus* dry matter production on the polluted soil to be 10t. The site contains about 21.8GBq of ^{226}Ra and the target post-remediation level is

0.2Bq /g DW soil. These assumptions allow an estimate for the number of crops necessary to reach the target level. The decrease in ^{226}Ra is illustrated for the two *Lupinus* species in Fig. 10, from which the number of years necessary to reduce the amount of ^{226}Ra at the site can be estimated as 144 for *albus* and 212 for *luteolus*.

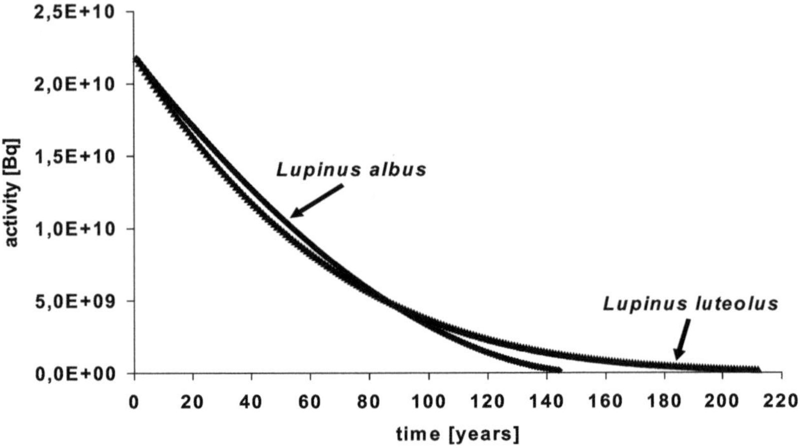

Figure 10. Log-log-linear uptake ratio.

6. Field studies – transfer of ^{226}Ra in plants

For this experiment, the accumulation of ^{226}Ra from contaminated soil at the K1 waste dump into the aerial parts of three pioneer woody plant species (*Alnus glutinosa, Betula pendula, Sambucus nigra*) was measured. The soil activity was 5 - 7Bq ^{226}Ra /g DW. The activity in the leaves of all three species increased during the experimental period. Leaf harvests were taken between May (soon after leaf appearance), when the average activity was about 0.10Bq/g DW, and October (at leaf fall; average activity about 0.20Bq/g DW). The highest activity was observed during August and September when woody plants are most metabolically and photosynthetically active. The highest activity was found in the leaves of birch (*Betula pendula*) at the end of the 2003 vegetation period (0.41Bq/g DW). The flowers and seeds of the trees tested contained less ^{226}Ra activity and seasonal changes were not significant except for birch (*Betula pendula*) during 2004. Significant increases in ^{226}Ra activity were seen in flowers and seeds, with the highest activity (0.42Bq/g DW) present at the end of the vegetation period. The activity in flowers and seeds was similar to that in leaves. Differences in the accumulation between the two vegetation seasons were clearly associated with variation in the annual rainfall (Fig. 11).

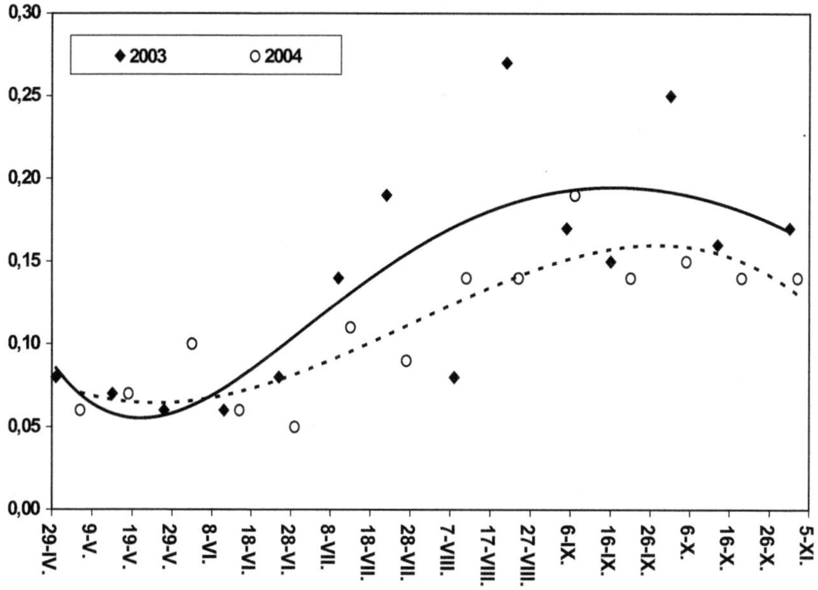

Figure 11. Transfer of ^{226}Ra nuclide from soil to alder tree (*Alnus glutinosa*) during the vegetation period. (Soudek et al., 2006c)

7. Constructed wetlands

Five gravel containers (8 to 16cm particle size) were each planted with one of the wetland plant species *Typha latifolia*, *Phragmites australis, Juncus inflexus, Carex buxbaumii* and *Iris pseudacorus*. Drainage water from two storage tanks of capacity $1m^3$ was circulated at 2 ml/min to the containers with *P. australis* and *T. latifolia*. Water flowed from the surface of the gravel to the bottom of container and back to the storage tank. The other three containers were filled by a single application. Decreases in the U concentration and ^{226}Ra activity over one month were determined (Fig. 12). The total amount of U in the input (765mg) and output of the containers with *Typha* and *Phragmites* were calculated. The plants and sludge from the containers were analysed and about 98.5% of the U was present in the sludge. Most of the U taken up by the plants was present in the roots (Table 3) (Soudek et al., 2005).

Table 3: Accumulation of uranium by *Phragmites* and *Typha* plants in constructed wetland.

		biomass [kg]	concentration [mg/kg]	amount of U [mg]
Phragmites australis	upper parts	0.36	1.1	0.396
	roots + rhizome	1.33	4.1	5.453
Typha latifolia	upper parts	0.24	0.6	0.144
	roots + rhizome	0.54	9.0	4.860
total		2.47		10.853

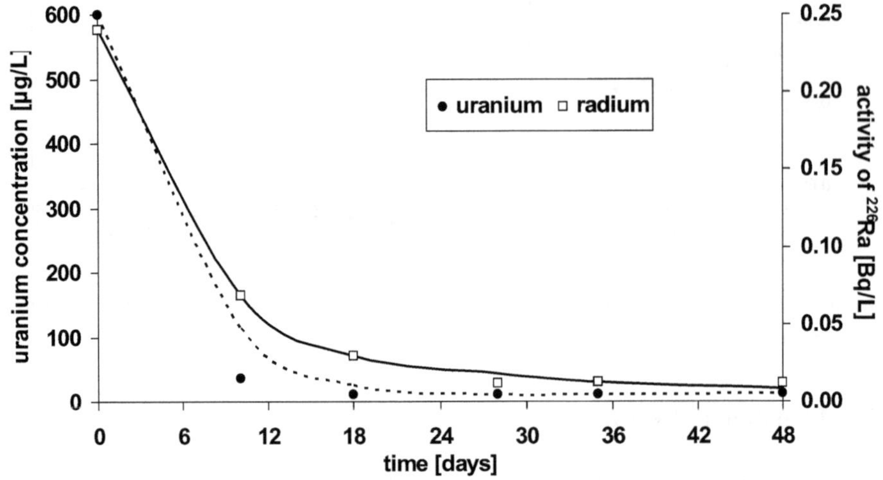

Figure 12. Decreasing of uranium concentration and ^{226}Ra activity in the drainage water during the circulation in constructed wetland.

8. Low-level contaminated sites

A large field experiment (about 0.7ha) was established in the area of a former spill, after the removal of bulk of the contamination. Hemp and flax cultivars were planted in May 2002 and 2003, and after harvest, the plant material was treated with the standard technological process for fibre production at Agritech Ltd. The activity of accumulated ^{226}Ra was determined in various parts of the plant. As all plants tested showed a level of activity indistinguishable from background, they were effectively non-radioactive (Table 4). Post-processing products similarly contained no ^{226}Ra activity (Table 5).

Table 4. Activity of flax and hemp cultivars after planting on low contaminated field.

Plant species	activity [Bq ^{226}Ra/g]
Linum usitatissimum "Jitka"	0.00
Linum usitatissimum "Atalante"	0.01
Cannabis sativa "Beniko"	0.02
Cannabis sativa "Juso"	0.02
Cannabis sativa "Silesia"	0.02

Table 5. Activity of different parts of flax cultivars after the standard technological process.

Flax parts after processing	cultivar "Atalante"	Cultivar "Jitka"
Contaminated soil	0.06 Bq/g of DW	
Seeds	0.000 Bq/g of DW	0.000 Bq/g of DW
Fibres	0.000 Bq/g of DW	0.000 Bq/g of DW
Awn	0.010 Bq/g of DW	0.000 Bq/g of DW
Deseeded capsule	0.005 Bq/g of DW	0.003 Bq/g of DW
Oil	0.000 Bq/mL	0.000 Bq/mL

9. Biomonitoring

Two biomonitoring experiments were performed. The first focussed on dump K3, which has been partially covered by inert and non-active material, and has been recultivated. High biomass producing crops (sunflower, corn, flax and hemp) were planted in 2003, but no ^{226}Ra activity was recovered from any of these plants. For the second investigation, in collaboration with the nuclear power plant NPP Temelín, eight test areas about 4 to 6km from the power plant were identified, and the level of background radiation (before the start-up of the reactor) was determined. The mean activity over the period 2001 – 2005 was 0.01 ± 0.001Bq/g DW, due to natural radionuclides (mainly ^{40}K), and no increase in activity due to the start-up of the NPP Temelín operation has been observed.

10. Conclusions

Our results show how radiophytoremediation can be applied to a practical situation, at least in the case of waste-water treatment, where the pattern of contaminant uptake is likely to resemble that observed in hydroponics experiments. For soil-cleaning purposes, the solubility of the contaminant(s) and its (their) soil mobility is the most limiting factor, along with the extent of the soil volume exploited by the roots of the remediating plant species.

Acknowledgements

This work was supported by COST project No. 859.10. The authors are grateful to M.Sc. M. Dvořáková for her final linguistic revision of the English text.

References

Arapis, G., Petrayev, E., Shagalova, E., Zhukova, O., Sokolik, G., and Ivanov, T., 1997, Effective migration velocity of ^{137}Cs and ^{90}Sr as a function of the type of soils in Belarus, *J. Environ. Radioac.* **34**:171–185.

Baker, A. J. M., and Brooks, R. R., 1989, Terrestrial higher plants which hyperaccumulate metallic elements – A review of their distribution, ecology and phytochemistry, *Biorecovery* **1**:81-126.

Belli, M., 1998, Long – Term Dynamics of Radionuclides in Semi – Natural Environments: Derivation of Parameters and Modeling., ed. M. Belli, Agenzia Nazionale per la Protezione dell'Ambiente, Rome.

Bennett, B., Bouville, A., Hall, P., Savkin, M., and Storm, H., 2000, Chernobyl accident: exposures and effects. In Proceeding of 10th IRPA International Congress, Hiroshima, Japan (14.-19. May 2000), T-12-1 (www.irpa.net/irpa10/cdrom/00713.pdf).

Broadley, M. R., Willey, N.J., and Mead, A., 1999, A method to assess taxonomic variation in shoot caesium concentration among flowering plants. *Environ. Pollut.* **106**:341-349.

Bruneton, J., 1995, Pharmacognosy, phytochemistry, medicinal plants. Intercept Ltd, Hampshire, England.

Bunzl, K., Schimmack, W., Belli, M., and Riccardi, M., 1997, Sequential extraction of fallout radiocesium from the soil: Small scale and large-scale spatial variability. *J. Radioanal. Nucl. Chem.* **226**:47–53.

Buysse, J., Van de Brande, K., and Meckx, R., 1995, The distribution of radiocaesium and potassium in spinach plants grown at different shoot temperatures. *J. Plant Physiol.* **146**:263-267.

Cline, J.F., and Hungate, F.P., 1960, Accumulation of potassium, caesium and rubidium in bean plants grown in nutrient solutions. *Plant Physiol.* **35**:826-829.

Cornish, J. E., Huddleston, G. J., and Levine, R. S., 1995, Phytoremediation of soils and water contaminated with toxic elements and radionuclides. In 7 ACS Special Symposium: Energing Technologies in Hazardous Waste Management, ed D.W. Tedder, American Chemical Society, Washington, DC, pp. 1352–1414.

Dushenkov, S., Mikheev, A., Prokhnevsky, A., Ruchko, M., and Sorochinsky, B., 1999, Phytoremediation of radiocaesium-contaminated soil in the vicinity of Chernobyl, Ukraine. *Environ. Sci. Technol.* **33**:469-475.

Entry, J.A., and Emmingham, W.H., 1995, Sequestration of ^{137}Cs and ^{90}Sr from soil by seedlings of *Eucalyptus tereticornis*. *Can. J. For. Res.* **25**:1044-1047.

EU publication, 1998, Atlas of caesium deposition on Europe after the Chernobyl accident, EUR 16733, Luxembourg.

Gauthier-Lafaye, F., 2002, 2 billion year old natural analogs for nuclear waste disposal: the natural nuclear fission reactors in Gabon (Africa). *C. R. Physiquée* **3**:839-849.

International Atomic Energy, 1989, Clean up of large areas contaminated as a result of a nuclear accident. Technical Reports Series No. 300.

Lasat, M.M., Fuhrmann, M., Ebbs, S.D., Cornish, J.E., and Kochian, L.V., 1998, Phytoremediation of radiocesium-contaminated soil: Evaluation of cesium-137 bioaccumulation in the shoots of three plant species. *J. Environ. Qual.* **27**:165-169.

Mengel, K., and Kirkby, E.A., 1978, Principles of plant nutrition. Ed. Int. Potash Institute.

Menzel, R.G., Heald, W.R., 1955, Distribution of potassium, rubidium, caesium, calcium and strontium within plants grown in nutrient solutions. *Soil Sci.* **78** (5):287-293.

Muramatsu, Y., Yoshida, S., and Bannai, T., 1995, "Tracer experiments on the behaviour of radioiodine in the soil-plant-atmosphere system", *J. Radioanal. Nucl. Chem. – Articles* **194** (2), 303-310.

Nepovím, A., Hebner, A., Soudek, P., Gerth, A., Thomas, H., Smrček, S., and Vaněk, T., 2005, Degradation of 2,4,6-trinitrotoluene by selected helophytes., *Chemosphere* **60**:1454-1461.

Nisbet, A. F., 1993, Effect of soil – based countermeasures on solid – liquid equilibria in agriculture soils contaminated with radiocaesium and radiostrontium. *Sci. Total Environ.* **137**:99–118.

Nishita, H., Steen, A.J., and Larson, K.H., 1958, Release of strontium-90 and cesium-137 from vina loam upon prolonged cropping. *Soil Sci.* **86**:195-201.

PhytoDec, 2004, A Decision Support System to quantify cost/benefit relationships of the use of vegetation in the management of heavy metal polluted soils and dredged sediments. (http://www.alterra-research.nl/pls/portal30/docs/folder/phytodec/fading_out.htm).

Procházka, S., Macháčková, I., Krekule, J., and Šebánek, J., 1998, Plant physiology. Academia Praha, pp. 484. (*in czech*).

Roca, M.C., and Vallejo, V.R., 1995, Effect of soil potassium and calcium on caesium and strontium uptake by plant roots. *J. Environ. Radioac.* **28**:141-159.

Sadolko, I., Arapis, G., and Davydchuk, V., 1995, Migration velocity of ^{137}Cs in the soils of Chernobyl area. *Journal of Radioecology* **3**, No.2, 7–13.

Sens, C., Scheidemann, P., Klunk, A., and Werner D., 1998, Distribution of ^{14}C -TNT and derivatives in different biochemical compartments of *Phaseolus vulgaris*. *Environ. Sci. Pollut. Res.* **4**:202-208.

Sens, C., Scheidemann, P., and Werner, D., 1999, The distribution of ^{14}C-TNT in different biochemical compartments of the monocotyledonous *Triticum aestivum*. *Environ. Pollut.* **104**:113-119.

Shaw, G., 1993, Blockade by fertilisers of caesium and strontium uptake into crops: effects on the root uptake process. *Sci. Total Environ.* **137**:119-133.

Smolders, E., and Shaw, G., 1995,. Changes in radiocaesium uptake and distribution in wheat during plant development: a solution culture study. *Plant Soil* **176**:1-6.

Soudek, P., Petřík, P., Vágner, M., Podracká, E., Tykva, R., and Vaněk, T., 2006a, Botanical survey and screening of plant species on contaminated soil of uranium waste depot., *European Journal of Soil Biology* (in press).

Soudek, P., Podracká, E., Vágner, M., Vaněk, T., Petřík, P., and Tykva R., 2004a, ^{226}Ra uptake from soils into different plant species. *J. Radioanal. Nucl. Chem.* **262**:187-189.

Soudek, P., Tykva, R., and Vaněk, T., 2004b, Laboratory analyses of ^{137}Cs uptake by sunflower, reed and poplar. *Chemosphere* **55**:1081-1087.

Soudek, P., Tykva, R., Vaňková, R., and Vaněk, T., 2006b, Accumulation of radioiodine from aqueous solution by hydroponically cultivated sunflower (*Helianthus annuus* L.). *Environ. Exp. Bot.* **57**:220-225.

Soudek, P., Valenová, Š., Benešová, D., Tykva, R., and Vaněk, T., 2006c, Transfer of ^{226}Ra nuclide from soil to plants during the vegetation period. *Journal of Environmental Radioactivity* (submited).

Soudek, P., Valenová, Š., Vavříková, Z., and Vaněk, T., 2006d, Study of ^{137}Cs and ^{90}Sr uptake by sunflower cultivated under hydroponic condition. *Journal of Environmental Radioactivity* **88**: 236-250.

Soudek, P., Vaněk, T., and Valenová, Š., 2005b, Method of removal of uranium and its decay series products from water. Patent application.

Tyson, M.J., Sheffield, E., and Callaghan, T.V., 1999, Uptake, allocation, accumulation and ecological implications of ^{85}Sr in bracken (*Pteridium aquilinum* L. Kuhn). *J. Environ. Radioac.* **46**:15-25.

U.S. Department of Energy Office of Environmental Management (DOEEM), 1998. Characteristics of important radionuclides. (http://web.em.doe.gov/idb97/tabb1.html).

Vaněk, T., Soudek, P., and Tykva, R., 2001. Study of radiophytoremediation, *Minerva Biotecnol.* **13**:177-121.

Velasco, R. H., Toso, J. P., Belli, M., and Sansone, U., 1997, Radiocesium in the northeastern part of Italy after the Chernobyl accident: Vertical soil transport and soil-to-plant transfer. *J. Environ. Radioac.* **37**:73–83.

von Fircks, Y., Rosén, K., and Sennerby-Forsse, L., 2002, Uptake and distribution of ^{137}Cs and ^{90}Sr in *Salix viminalis* plants. *J. Environ. Radioac.* **63**:1-14.

Watt, N.R., Willey, N.J., Hall, S.C., and Cobb, A., 2002,. Phytoextraction of 137Cs: the effect of soil ^{137}Cs concentration on ^{137}Cs uptake by *Beta vulgaris*. *Acta Biotechnol.* **22**:183-188.

White, P.J., Swarup, K., Escobar-Gutierrez, A.J., Bowen, H.C., Willey, N.J., and Broadley, M.R., 2002, Selecting plants to minimise radiocaesium in the food chain. *Plant Soil* **249**, 177-186.

Willey, N., Hall, S., and Mudigant, A., 2001, Assessing the potential of phytoremediation at a site in the U.K. contaminated with ^{137}Cs. *Int. J. Phytorem.* **3**:321-334.

Wolterbeek, H. T., and van der Meer, A. J. G. M., 1996, A sensitive and quantitative biosensing method for the determination of γ - ray emitting radionuclides in surface water. *J. Environ. Radioac.* **33**:237–254.

Zhu, Y.G., Shaw, G., Nisbet, A.F., and Willkins, B.T., 1999, Effect of external potassium supply on compartmentation and flux characteristics of radiocaesium in intact spring wheat roots. *Ann. Bot.* **84**:639-644.

PHYTOREMEDIATION – SOME CASE STUDIES CONDUCTED AT WAU

STANISLAW W. GAWRONSKI* AND HELENA GAWRONSKA
Laboratory for Basic Sciences in Horticulture, Faculty of Horticulture and Landscape Architecture, Warsaw Agricultural University, Nowoursynowska 159, 02-776 Warsaw, Poland

*To whom correspondence should be addressed.

Abstract- We present an overview of the methodology for bioassays and other techniques used in phytoremediation studies under controlled conditions, as practised at the Laboratory for Basic Sciences in Horticulture of Warsaw Agricultural University. These relate both to urban and industrial areas and concentrate on the major pollutants heavy metals (HMs), polycyclic aromatic carbohydrates (PAHs) and de-icing salt (NaCl). Depending on the specific goals of particular experiments, the following bioassays are employed: seed germinability, seedling vigour, assessment of biomass production and partitioning, and accumulation and distribution of pollutants. In studies of the physiological, biochemical and molecular plant responses to pollutants, plant gas exchange is measured by infrared gas analysis, plant water status by a dew point voltmeter (assessing water and osmotic potentials and relative water content), wax deposition by chloroform washing, changes in total glutathione and low molecular weight organic acid levels by HPLC, the expression of genes encoding metallothionein by PCR and DNA sequencing, and gene expression profiling by micro-array/DNA chip). Some of these will be presented briefly, together with the results of some case studies.

Keywords: heavy metals, PAHs, salt, germination biotest, seedling vigour, gas exchange, water status, wax deposition, gene expression, total glutathione, organic acids, HPLC

N. Marmiroli et al. (eds.), Advanced Science and Technology for Biological Decontamination of Sites Affected by Chemical and Radiological Nuclear Agents, 159–175.

1. Introduction

Studies conducted at WAU are focused on: (i) the characterization of the pollution level of selected urban and industrial areas, (ii) the evaluation of pollutant effect on the soil environment and vegetation, (iii) the selection of species/cultivars tolerant to pollutants, (iv) the evaluation of pollutant uptake and accumulation in various organs with special attention paid to easily harvestable organs, (v) the degradation/detoxification of pollutants by plants, (vi) the elaboration of recommendations for plant cultivation in polluted urban sites, and (vii) the physiological, biochemical and molecular mechanisms involved in plant acclimation to pollution. Various methodological approaches, bioassays and techniques are employed to these ends. For the preliminary screening of a large number of species/cultivars and/or testing responses to a wide range of pollutant type and concentration, biotests must be simple, rapid, easy to perform and highly replicable. Sophisticated methods are necessary to unravel the mode of action of particular pollutants as well as the mechanisms of the plant response and its tolerance. Seeds, seedlings, fully developed annual and perennial herbaceous plants and rooted shrub and tree cuttings, as well as plant parts and extracts are targeted.

2. Biotests

For a preliminary screening of a large number of species/cultivars and/or for the testing of responses to a wide range of pollutants, especially when several concentrations of each pollutant need to be evaluated, germinability tests and seedling vigour tests are highly suitable. Germinability tests the capacity and dynamics of the germination process, while seedling vigour tests can provide a more detailed set of information regarding tolerance to a given pollutant, based on seedling survival, length, weight, and organ malformation. The most commonly employed tester species are lettuce (*Lactuca sativa* L.), mustard *(Sinapis alba* L.), cucumber (*Cucumis sativum* L.), and *Lepdium sativum,* but for some specific purposes other species are also used. Seeds are generally germinated in Petri dishes, layered with two filter paper discs moistened with distilled water containing a known concentration of pollutant, and with water only as control. The volume of solution must be sufficient to provide adequate supply according to the number of seeds per dish, seed size, dish size and germination period; for 9cm diameter dishes, 5-10ml is normal. Seed number is typically 25-100 per replication (Petri dish) and experiments usually consist of three to six replicates. Generally at least five replicates are necessary to achieve reproducible results. Seeds are evenly distributed in the dish and incubated for 7-14 days in the dark at an optimal temperature. For the determination of only germination capacity, the number of germinated seeds is counted at the end of the experiment, but if the dynamics of germination are important, then a daily count is taken, and germinated seeds are removed. In some laboratories, seeds and seedlings are cultured on an agar medium supplemented with nutrients and pollutants (Heidenreich et al. 2001). Where the number of treatment combinations is too high to fit in a single run, then controls must be included for each set of dishes. For the evaluation of seedling vigour, seeds can be placed between two rolled up strips of appropriately moistened filter paper moistened with appropriate solutions, then placed in covered jars containing the same culture solution (Perry 1987, Gawronska and Grzelak 1993). Seedling vigour experiments are generally completed before germinability ones (for instance, seven, as opposed to 14 days, for mustard). After about seven days, the length, fresh and dry weights of hypocotyls/coleoptiles and root(s) are obtained, along with any observations regarding seedling malformation. A typical example of a biotest result, measuring the germination of some amaranth cultivars germinated under salt stress is shown in Figure 1.

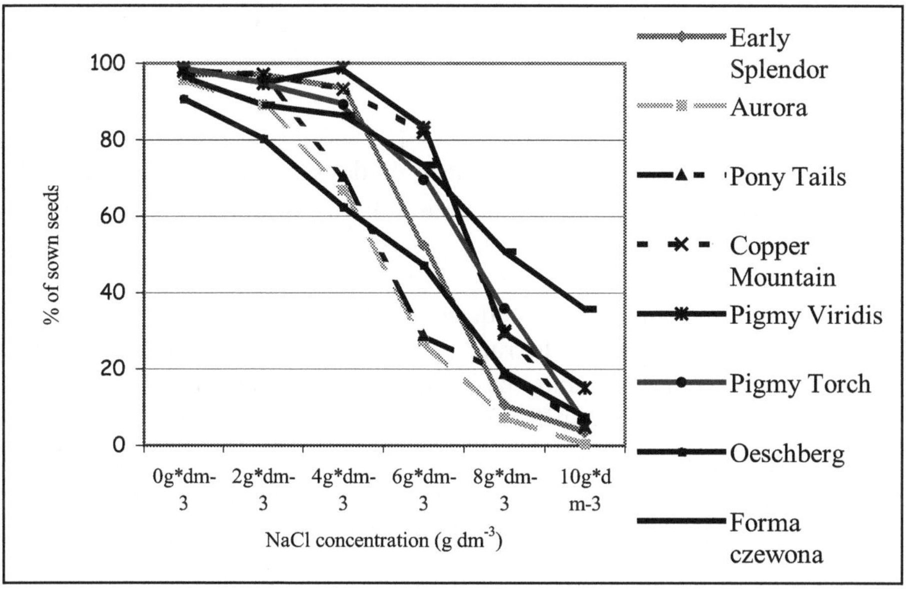

Figure 1. Germination of selected cvs. of amaranth as affected by various concentrations of de-icing salt.

From germination and seedling growth data, the following indices can be calculated:

11. 1) Tolerance index (Wilkins, 1957), according to the formula:

TI (%) = length of hypocotyls in heavy metal/ length of hypocotyls in control x 100

(The length of other plant organs, or the level of germination can also be used.)

12. 2) LD_{50}, LD_{80}, representing the dose at which 50 or 80% of seed will not germinate, or of seedlings which will not survive beyond a specific period of time.
13. 3) Index Pippera (PI) (Pipper, 1952), used for the evaluation of germination dynamics where data are collected daily. PI reflects to the number of days needed for one seed to germinate and is calculated according to the formula:

$$PI = \frac{x_1 s_1 + x_2 s_2 + \ldots + x_n s_n}{s_1 + s_2 + \ldots + s_n}$$

where x_1, x_2,....x_n represent the number of days for germination, s is the number of seeds that germinate on a particular day, and n is the last day of the experiment.

These biotests are simple, require minimal input of labour, produce rapid results and are cheap to run. As a result they can be deployed in almost any laboratory, and have no requirement for sophisticated equipment. However, they suffer from a number of limitations:

- not all plant species are propagated from seeds; for species recommended for phytoremediation, vegetative propagation is preferable, because it avoids problems of genetic instability.
- tolerance to pollutant determined at the germination stage often does not correspond with behaviour at later growth stages, when phytoremediation would take place; seeds are generally more tolerant than growing plants.

- no data are produced which describe plant growth, biomass production and partitioning, or pollutant uptake, translocation and distribution, all of which are important for effective phytoremediation.
- the physiological, biochemical and molecular basis for tolerance to pollutants is more properly studied at the autotrophic growth stage, when the photosynthetic apparatus is fully developed; most pollutants affect photosynthesis with respect to light harvesting, electron transport and the enzymatic conversion of carbon dioxide.

Germinating seeds and seedlings can also be used for physiological, anatomical, cytological and molecular studies, but these require specific methodology which is neither simple nor fast.

3. Techniques and methods used for physiological studies at the autotrophic stage

Among most frequently studied physiological responses to environmental pollution at the autotrophic stage are biomass production and partitioning, plant gas exchange, plant water status, membrane integrity and changes in plant stress hormones levels. For such studies, plants are grown either in pots with peat moss supplemented with nutrients, or in aerated hydroponic culture with Hoagland's nutrient solution (Arnon and Hoagland, 1940). Plants at the autotrophic stage (with fully developed photosynthetic apparatus) are exposed to pollutants such as heavy metals or de-icing salt by their inclusion in the growing medium. For pot-grown plants, they can be applied in a controlled manner, either via watering or when transplanting to new pots. For hydroponic-grown plants, they are added directly to the nutrient solution, most often at weekly intervals. Sufficient aeration of the nutrient solution is necessary to avoid anaerobiosis. This can be achieved either by continuous aeration or by pulsing for at least six hours per day. The quantity of nutrients and the time of exposure to pollutants are adapted to suit the specific species, plant age and experimental design requirements. Hydroponic culture is of especial importance to phytoremediation studies, because it ensures easy access to the root system.

Biomass accumulation and its distribution are usually determined after a prolonged period of exposure to pollutants. At harvest, the plant is divided into its constitutive organs, the fresh weight of each is recorded and the material is then oven-dried for 1-2 h at 105°C and at 70°C for further 2-3 days, preferably using an air circulation system. Drying time can vary depending on type of tissue.

For **plant gas exchange** measurements, the most frequently used method is infrared gas analysis (IRGA), based on the absorption of infrared irradiation by CO_2 and H_2O. Changes in the concentration of CO_2, due to photosynthesis or respiration, and in H_2O, due to transpiration, are monitored and calculated per unit time and area or biomass. An advantage of this method is that time course measurements can be performed on the same leaves because the sampling is non-destructive. Several types of plant gas exchange apparatus are currently commercially available, including portable versions. They are typically automated and computerized, and allow many measurements to be collected in a single field trip (but are also useful for laboratory, greenhouse and growth chamber experiments). Simultaneous measurements, integrated over seconds to a few minutes, are made of the intensity of photosynthesis and transpiration, the stomatal resistance, the internal CO_2 concentration, organ temperature and incident photosynthetically active radiation (PAR). Figure 2 illustrates changes in photosynthesis (A), transpiration (B) and stomatal resistance(C) of hydroponically grown *Arabidopsis thaliana* L. plants exposed to Cd^{2+}, Cu^{2+} and Pb^{2+} as measured with a LICOR 6200 Photosynthesis System based on IRGA.

To measure the **dark respiration rates** of photosynthetic tissue, measurements have to be performed in complete darkness, or with apparatus specially designed for this purpose. Long term rates of photosynthesis can be evaluated indirectly by dry matter accumulation. However, this is a somewhat inaccurate measure, because the measured amount of dry matter accumulated under stress conditions is less than it would be under non-stressed ones.

Plant water status is affected by environmental pollution and consequently influences plant function at every level of biological organisation. It can be characterized by measurements of the relative water content (RWC), the water deficit, the water potential (Ψ_w) and the osmotic potential (Ψ_o), along with transpiration rate and stomatal resistance. Since for the latter four parameters, tissue samples are removed from the plant, they are usually determined in the end of an experiment. If several sampling times are needed, then additional plants/replicates must be included.

Transpiration rate and stomatal resistance are the most commonly measured plant gas exchange parameters using IRGA. Their behaviour in *A. thaliana* plants exposed to heavy metal ion stress is demonstrated in Figure 2 B and C.

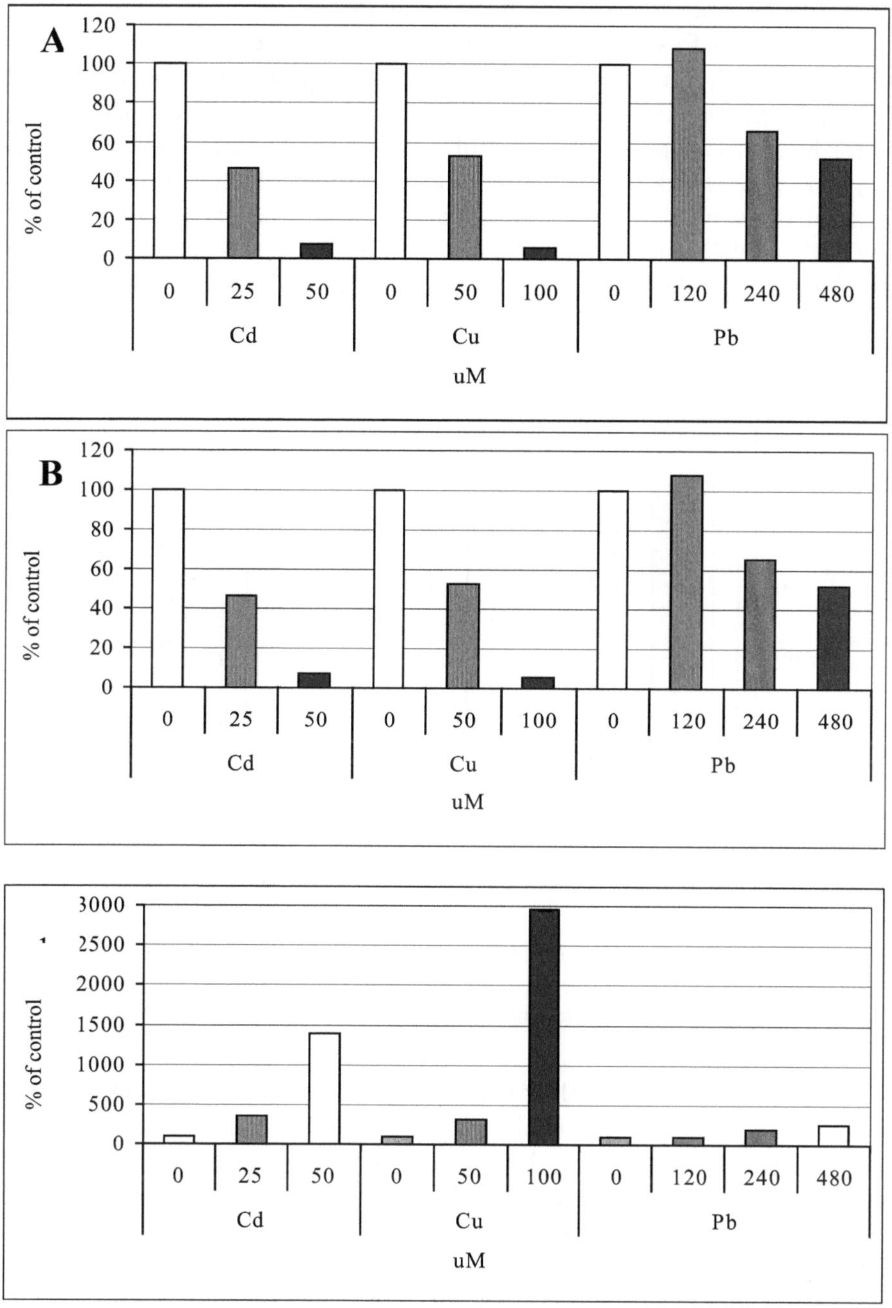

Figure 2. Intensity of photosynthesis (A), transpiration (B) and stomata resistance (C) of *Arabidopsis thaliana* L. plants exposed to Cd^{+2}, Cu^{+2} and Pb^{+2} ions in hydroponics culture for 22 and 29 days. Measurements were made using IRGA method with Photosynthesis System LICOR 6200, Lincoln, Nebraska. Data are expressed as % of control ± SE, n=5.

RWC can be determined in any laboratory, as it only requires the measurement of fresh and dry weights. This parameter reflects the level of plant or organ hydration (i.e., whether or not they are suffering from water stress). Plant material is weighed (fresh weight, FWa) and then held at 100% humidity. Upon reaching full turgor (usually after several hours), they are reweighed (fresh weight at full turgor, FWt). Thereafter, the material is oven-dried until a constant dry weight is reached (dry weight, DW). The RWC can then be calculated as [(FWa – DW)/ (FWt – DW)] x 100 (Bray et al.2000). Changes in RWC as affected by Cu^{2+}, Cd^{2+} and Pb^{2+} stress are shown in Figure 3.

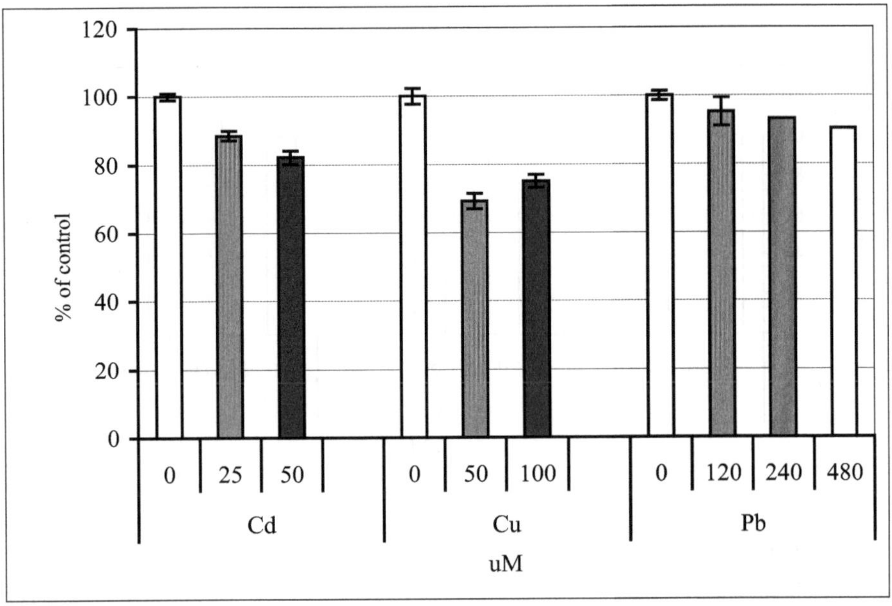

Figure 3. Relative water content in leaves of *Arabidopsis thaliana* L. plants exposed to Cd^{+2}, Cu^{+2} and Pb^{+2} ions in hydroponics culture for 29 days. Data are expressed as % of control ± SE, n=5.

The two remaining of plant water status parameters (water and osmotic potential) reflect free water availability. These can be measured with a Dew Point Microvoltmeter (e.g., the WESCOR HR 33T [Logan, Utah]), but other methods and equipment can be applied. Osmotic potential can be determined from either fresh or frozen plant samples, but water potential requires fresh tissue. The water potential of leaf discs, strips or roots of the same size/weight can be measured with the WESCOR device (for detailed methods see http://www.wescor.com/environmental/index.phtml). Osmotic potentials are measured by moistening filter paper discs with the liquid squeezed from sample tissues.

Proper cell/organelle function is strongly affected by various stresses including pollutants, and relates to the level of free water available in the cell. When the level of free water falls below a certain (species-dependent) threshold, injuries to the cell membranes occur, and membrane function is disrupted. The first symptom is often the loss of selective membrane permeability, leading to electrolyte leakage, which can be measured conductometrically. Control and treated samples (of known fresh weight or numbers of leaf discs) are washed with distilled water (to remove all ions from cut cells), held for 2–3h in a fixed volume of distilled water before measuring the electroconductivity of the medium. After an incubation at 95-100°C for 10min, the material is cooled to room temperature, the volume of liquid is restored to its original level, and the electroconductivity is re-measured (treated T_2, control C_2). Membrane injury is then calculated by the formula:

$$\% \text{ of injury} = 1- [1-(T1/T2)/1-(C1/C2)] \times 100$$

We have studied changes in membrane integrity in *A. thaliana* plants exposed to selected concentrations of Cd^{2+}, Cu^{2+} and Pb^{2+} (Figure 4).

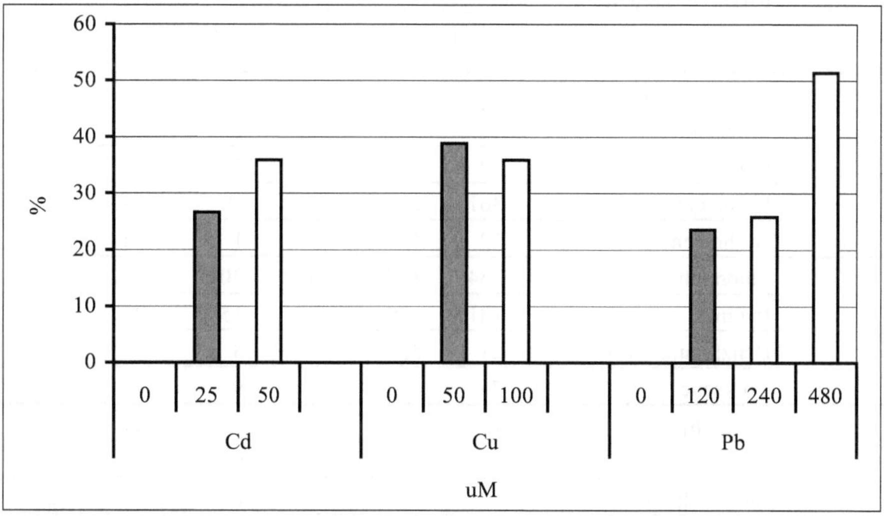

Figura 4. Membrane injury in leaves of *Arabidopsis thaliana* L. plants exposed to Cd^{+2}, Cu^{+2} and Pb^{+2} ions in hydroponics culture for 29 days. Data are mean of 5 replications expressed as % of injury recorded in control.

Polycyclic aromatic hydrocarbons (PAHs) are hydrophobic compounds which are absorbed by the cuticular wax layer, which acts as a trap for these pollutants. Plants experiencing PAH exposure (for example, on the verges of roads carrying heavy traffic) often show an increased level of wax deposition on their aerial parts. The amount of wax on the leaf and stem surfaces can be measured, according to Lurrie et al. (1996 and references therein) after two washes with chloroform. The amount of chloroform used and the washing time differ and need to be adjusted to each specific plant species in pilot experiments. We have used about 25ml chloroform per 8g leaf fresh weight. The first wash occupies between 20-40s, and the second is about 50% shorter. The two washes are combined and filtered through glass or filter paper, the chloroform is evaporated, and residue is weighed and expressed in the form of $\mu g.cm^{-2}$ of leaf area. We have found that the amount of wax deposited on the leaf surface of *Robinia pseudoacacia* is inversely proportional to the distance from the road, falling from $50.8\mu g.cm^{-2}$ at 1-3m from the road to $24.4\mu g.cm^{-2}$ wax at over 100m (Gawronski and Raczka 2004). The quantity of absorbed PAH can be determined with mass spectrometry, and in *R. pseudoacacia,* 17 different PAHs were found, of which seven are recognized carcinogens (Table 1) (Gawronski and Raczka 2004).

Table 1. Concentration of PAHs (mg kg^{-1} DM) in the leaves and in autumn and spring stems of *Robinia pseudoacacia* collected at two distances from heavy traffic.

N° of rings	PAHs compound	Leaves		Stem in autumn		Stem in spring	
		1-3m	>100m	1-3m	>100m	1-3m	>100m
3	Acenaphtylene	21	6	2	2	8	2
	Acenaphtene	41	5	2	2	4	2
	Fluorene	39	26	4	3	14	5
	Phenanthene	361	202	48	30	235	97
	Anthracene	19	12	3	1	14	2
4	Fluoranthene	264	172	41	20	332	52
	Pyrene	181	100	37	15	257	33
	Benz[a]anthracene	21	33	11	3	19	3
	Chrysene	76	112	26	9	120	29
5	Benzo[b]fluoranthene	19	32	9	5	29	8
	Benzo[k]fluoranthene	8	12	3	t.a	9	t.a*
	Benzo[e]pyrene	32	15	4	t.a	17	t.a
	Benzo[a]pyrene	6	9	t.a	t.a	7	t.a
	Perylene	9	t.a	t.a	t.a	t.a	t.a
	Dibenz[a,h]anthracene	t.a	t.a	t.a	t.a	7	6
6	Indeno[1,2,3c,d]pyrene	9	12	t.a	t.a	14	t.a
	Benzo[g,h,i]perylene	10	12	6	t.a	17	6
	Σ	1116	760	196	90	1105	245
	Cancerogenic	162	213	53	177	208	46

*t. a. - trace amount

Plants exposed to heavy metal stress often increase the pH of the growing medium in order to reduce their uptake. We have recorded such a response in *Sinapis alba* L cv. Nakielska plants grown in the presence of Pb^{2+} (Figure 5) (Wińska-Krysiak and Gawroński, 2002). This parameter

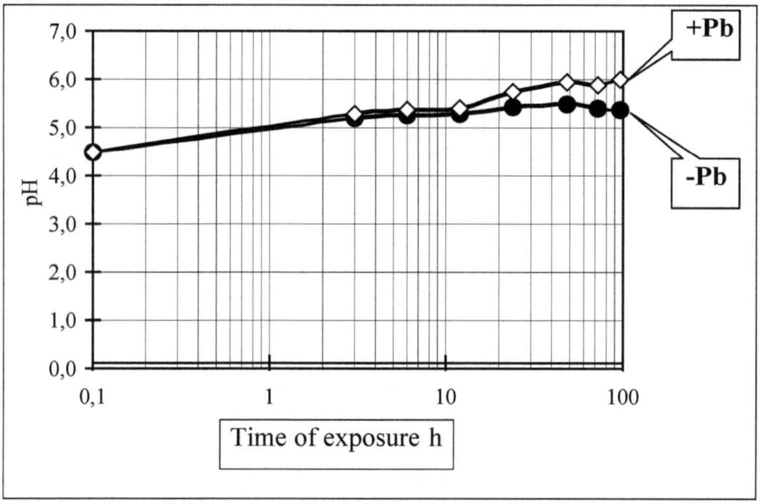

Figure 5. Changes in pH of hydroponics solution by *Sinapis alba* L. plants exposed to Pb^{+2} ions in concentration of 25 ppm.

can be measured in any laboratory, as only a pH meter is needed. Some plant species excrete more acidic compounds (low molecular weight organic acids) when grown in the presence of heavy metals. These can be detected by HPLC. Thus, for example, Raczka (2004) reported increased amounts of malic acid in the hydroponic solution when *Amorpha fruticosa* was exposed to 45ppm Pb^{2+} (Figure 6). As a result, the nutrient pH was reduced, in contrast to the behaviour of *S. alba* mentioned above.

Exposure to pollutants generates changes in the profile of gene expression. Understanding the full response to stress requires the study of all levels of biological organisation. The sequencing of the whole genomes of several plant species has opened the possibility of applying functional genomic analysis, and in particular microarray technology, to profiling gene expression. This approach allows for the simultaneous measurement of expression levels of thousands of genes in a single experiment. Micro-array methodology is rapidly being miniaturized and automated, and this development will help increase its accessibility to the research community. At present, however, it remains a relatively expensive technique. The microarray method exploits preferential binding of labelled single stranded nucleic acid sequences of the tested and control samples to complementary

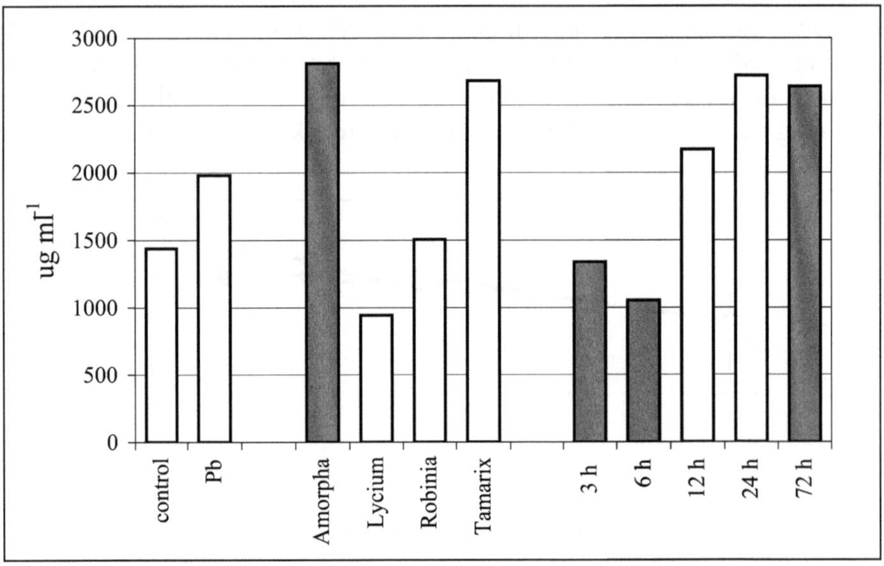

Figure 6. Concentration of malic acid (mg ml^{-1}) exuded to growing medium as influenced by species and exposure time to Pb^{+2} in concentration of 45 ppm.

sequences of DNA immobilized on the array. A typical micro-array experiment consists of the preparation of either cDNA probes (ESTs, PCR products, cloned genes, full cDNA sequences) or oligonucleotides; their immobilization to a matrix at predetermined locations; the isolation and labelling of RNA from test and control samples; the hybridisation of labelled RNA to the array; scanning, image processing, data normalization and analysis; and finally the derivation of function of differentially expressed sequences.

Here we provide, as an example, data from a Affymetrix chip micro-array gene expression study of the effect of Pb^{2+} stress on *A. thaliana* (Gawronska et al. 2004). The chip carries 8200 genes (about 30% of the full genome). About 50% of these genes were differentially expressed, with some being induced, and others suppressed (Table 2), and with a further group showing changed levels of expression (Table 3). These genes were classified to various categories (Figure 7). More than half are of unknown function. Only 342 of the differentially expressed genes are known to be directly related to the stress response. Of particular interest to phytoremediation are the 40 genes encoding metallothioneins and enzymes involved in the biosynthesis of phytochelatins, as well as the 138 genes encoding ion channels, as these relate closely to plant acclimation to heavy metals stress. Further validation of these candidate genes is, however, required.

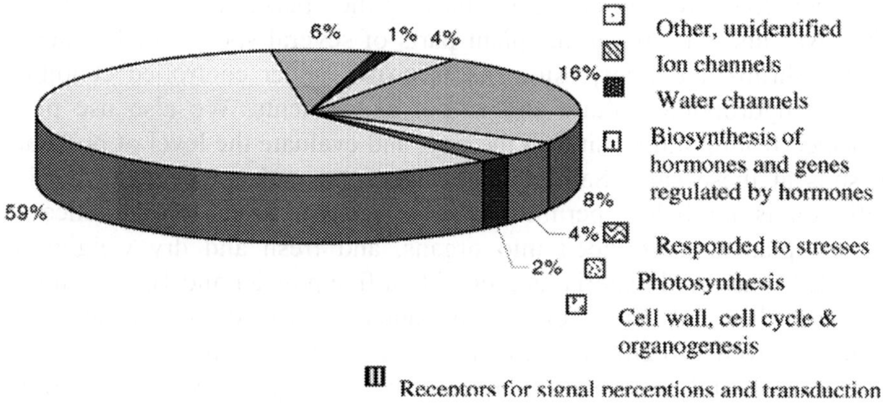

Figure 7. Categories of genes differentially expressed in *Arabidopsis thaliana* L. plants exposed to 240 μM of Pb^{+2} ions in the hydroponics culture for 14 days.

Table 2. Number of genes induced (P) or suppressed (A) in *Arabidopsis thaliana* plants exposed to 240 μM of Pb^{+2} ions for 14 days.

Organ	Induced		Suppressed	
	Control = A 25=P	Control = A 50=P	Control = A 25=P	Control = A 50=P
Leaves	118	256	524	287
Roots	147	241	504	580

Table 3. Number of genes with increased or lowered levels of expression in *Arabidopsis thaliana* plants exposed to 240 μM of Pb^{+2} ions for 14 days.

Organ	Lower level of expression			Higher level of expression		
	120 vs. control	240 vs. control	120 vs. 240	120 vs. control	240 vs. control	120 vs. 240
Leaves (L)	658	889	689	598	1231	959
Roots (R)	1084	1398	1088	877	1203	1196
Common (L+R)	142	284	132	85	229	222
Total	1600	2003	1645	1390	2205	1930

From the phytoremediation point of view, an understanding of the mechanisms underlying pollutant uptake, translocation to harvestable organs and/or their detoxification is critical for developing and elaborating the technology. We are currently studying the uptake and accumulation of Pb^{2+}, Na^{+} and Cl^{-}, in various plant parts of several species of herbaceous plants, shrubs and trees, using plants grown under controlled conditions both in hydroponic culture and in pot experiments. We also use plants collected from polluted sites in the city, and evaluate the level of pollutants in soils at these sites. Samples collected from parks, where the level of pollution is far below permitted thresholds, are taken for reference. At harvest plants, are divided into organs, and fresh and dry weights are recorded. The dried material is ground to a fine powder and 1g sub-samples used for the analysis of elemental content. Roadside soil samples are collected from 0-0.2m, at various distances (0.3- 6.0m) from the road's edge. Soil salinity and pH are assessed by soaking in distilled water, and measurement by electroconductometer/pH meter. The content of Na^{+} and Pb^{2+} is determined after mineralization at 450°C, followed by dilution in 0.5N HCl and finally made up to the required volume with de-ionised water. A flame or graphite (depending on the concentrations of Pb^{2+}) atomic absorption spectrophotometer is used for the quantification of cation content. Cl^{-} content is determined by titration. These analyses have shown that the de-icing of roads has increased both the soil salt concentration and pH up to 6m from the road's edge. Salt concentration ranged from 0.81-5.91mS and the pH from 6.22-8.68, with the higher salt concentrations and pH present closest to the road. Increased salinity along highways has resulted in the disappearance of previously common grasses, and their replacement by successions of halophytes, such as *Atriplex tataricum*. De-icing highways during the winter harms neighbouring vegetation not only due to changes induced in the soil environment, but also because of dispersion of salty water by heavy traffic. We have noted significant damage to trees and shrubs along the highways. The concentrations of selected elements in the above ground plant parts of three cultivars of amaranth as affected by salinity levels in growing medium are shown in Table 4. Na^{+} and Cl^{-} accumulated in cultivars in a concentration-dependent manner, but there appear to be significant genotypic differences between them. The Pb^{2+} uptake and accumulation in various organs of several cultivars of *Canna x generalis* plants is shown in Table 5, showing that the ion is not only very efficiently taken up, but also that a high proportion is accumulated in the easily harvestable rhizomes (especially for the cultivars Delibab and Halina). Given the high biomass potential of this species, and its decorative value, it seems to represent a promising candidate for the phytoremediation of polluted urban sites.

Table 4. The effect of salinity in growing medium on accumulation of Na^{+}, Cl^{-}, K^{+}, Ca^{2+}, Mg^{2+} in dry matter of plants of *A. paniculatus* cv. Copper Moutain i Monarch, *A. caudatus* cv. Pony Tails. Data are mean ±SE, of 2 vegetations seasons with 3 replicates in each.

Species/cultivar	Salt concentration (g NaCl·dm^{-3})	Ions concentration in dry matter (mg·g^{-1} d.m.)					Ratio
		Na^{+}	Cl^{-}	K^{+}	Ca^{2+}	Mg^{2+}	K/Na
Amaranthus	2	0,433	32,97	37,91	18,58	4,85	87,55
paniculatus	4	0,475	49,38	49,97	25,61	6,37	105,2
cv. Copper	8	14,29	79,19	52,21	30,99	6,37	3,65
Moutain	16	64,97	158,1	49,90	36,84	5,71	0,77
Amaranthus	2	0,209	27,83	47,45	15,59	4,44	227,03
caudatus cv.	4	0,300	33,27	54,15	19,93	5,54	180,5
Pony Tails	8	18,66	70,95	53,37	30,64	6,15	2,86
	16	76,14	170,83	55,31	31,61	4,45	0,73
Amaranthus	2	0,357	19,53	51,32	18,47	3,63	143,75
paniculatus	4	0,895	25,00	50,55	26,23	4,54	56,48
cv. Monarch	8	33,38	70,23	50,29	29,56	4,19	1,51
	16	85,05	159,3	57,90	35,24	4,23	0,68

$NIR_{Na+} = 2,58$	$LSD_{Na+} = 2,58$
$NIR_{Cl-} = 5,71$	$LSD_{Cl-} = 5,71$
$NIR_{K+} = 2,71$	$LSD_{K+} = 2,71$
$NIR_{Ca2+} = 1,07$	$LSD_{Ca2+} = 1,07$
$NIR_{Mg2+} = 0,25$	$LSD_{Mg2+} = 0,25$

The above examples have shown that laboratory studies carried out under controlled conditions are important for gaining an understanding of the processes and mechanisms involved in plant acclimation to environmental stresses. The results of these studies, besides being of academic interest, can also inform the development of phytoremediation technology. We anticipate seeing the cultivation of plants not only for aesthetic and landscape architecture reasons, but also for the rescue of polluted sites.

Table 5. Lead content and concentrations in organs and whole plants in selected *Canna x generalis* cultivars treated with 45 ppm of lead.

Canna genotype	Pb in plant organs				Pb in whole plants	
	Organs	Content in µg	Concentration in µg g^{-1} DM	In %	Content in µg	Concentration in µg g^{-1} DM
President	Roots	9486,4	45710,0	83,4	11299,0	4184,8
	Rhizome	16668,2	1112,1	15,2		
	Leaves	144,4	144,4	1,3		
Alberich	Roots	6166,0	30830,0	55,3	10954,0	2235,5
	Rhizome	4745,8	2063,4	44,3		
	Leaves	42,2	17,6	0,4		
Herkules	Roots	7662,3	25541,0	71,3	10743,0	1451,7
	Rhizome	2950,3	1282,7	27,5		
	Leaves	130,4	27,1	1,2		
Delibab	Roots	0,0	0,0	0,0	10721,0	3371,3
	Rhizome	10556,1	5060,1	98,5		
	Leaves	130,4	149,9	1,5		
Halina	Roots	0,0	0,0	0,0	2876,0	586,9
	Rhizome	2545,8	1106,8	88,5		
	Leaves	330,2	127,3	11,5		

Acknowledgements

The authors express their thanks to M. Ciurzynska and M. Wrochna for providing some results of their Ph.D. studies to illustrate the methods and techniques described here.

References

Arnon D. I., and Hoagland, D. R., 1940, Crop production in artificial solutions and soils with special references to factor influencing yield and absorption of inorganic nutrients, *Soil Sci.* **50**:463-471.

Bray, E., Bailey-Serres, J., and Weretilnyk, E., 2002, Response to abotic stresses. In: Buchanan B., Gruissem R., Jones R. eds. Biochemistry and Molecular Biology of Plants. American Society of Plant Physiologist, Rockville Maryland, USA. pp. 1162.

Gawronska, H., and Grzelak, K., 1993, Seed germination and seedling vigourur of triticale (x Triticosecale Wittmack) under drought stress. *Plant Varieties and seeds* **6**:9-19.

Gawrońska, H., Ciurzyńska, M., Wiench, M., and Gawroński, S. W., 2004, Changes in profile gene expression in *Arabidopsis* induced by Pb^{+2} ions. Abstracts of 14th Congress of Federation of European Societies of Plant Biologists. Cracow, Poland, p.221.

Gawronski, S. W., and Raczka, M., 2004, Lead uptake and PAHs capture by plants in urban areas, Abstracts of COST Action 859. Working Group 4 Meeting. Integration and application of phytotechnologies. October 28-20, Leipzig, Germany. p.28.

Heidenreich, B., Mayer, K., Sandermann, J. R., and Ernst, D., 2001, Mercury –induced genes in *Arabidopsis thaliana*: identification of induced genes upon long-term mercuric ion exposure, *Plant Cell Environ* **24**:1227-1234.

Lurrie, S., Fallik, E., and Klein, J. D., 1996, The effect of heat treatment on apple epiculticular wax and calcium uptake, *Postharvest Biology and Technology* **8**:271-277.

Perry, D. A., 1987, Handbook of vigourur testing methods. ISTA, Zurich.

Pipper, H., 1952, Das Saatgut. (Ed. P. Parey) pp. 271. Berlin, Germany: Paul Parey

Raczka, M., 2004, Ocena przydatności wybranych gatunków drzew i krzewów w fitoremediacji terenów zurbanizowanych. [Evaluation of the usefulness of selected trees and shrubs in phytoremediation of urban areas] Ph.D. thesis. Warsaw Agricultural University.

Wilkins D.A., 1957, A technique for the measurement of lead tolerance in plants, *Nature* **180**:37-38.

Wińska-Krysiak, M., and Gawroński, S. W., 2002, Fizjologiczney aspekt tolerancji i hiperakumulacji ołowiu w wybranych genotypach Brassica. [Physiological aspects of lead tolerance and hyperaccumulation in Brassica species.] Advances of Agricultural Sciences Problem Issues. Ecophysiological aspects of plant responses to stress factors. Polish Academy of Sciences. 481, part II, 605-613.

MANAGEMENT OF PASSIVE BIOLOGICAL WATER TREATMENT SYSTEMS FOR MINE EFFLUENTS

CHRISTIAN KUNZE*, GUNTER KIEßIG, ANNETTE KÜCHLER
WISUTEC Wismut Umwelttechnik GmbH, Jagdschänkenstr., D-09117 Chemnitz, Germany, info@wisutec.de

*To whom correspondence should be addressed.

Abstract- This contribution describes the approach of WISUTEC, a subsidiary of the mine remediation company WISMUT, to passive biological water treatment as a long-term solution to water contamination problems at former uranium mining and milling sites. We introduce the WISMUT project, with special emphasis on water management and treatment at site closure, and the subsequent remediation process of decommissioned mining and milling sites, and discuss a variety of water-related problems within this framework. A peculiarity of mining and milling operations in Germany, as in Europe in general, compared to other typical mining regions worldwide, is the relatively high population density in the affected mining areas and the relative scarcity of land. As a consequence, strict requirements with respect to technical solutions are applied to mine closures, in particular relating to restrictions on the land surface available for semi-natural and constructed wetlands. The regulatory expectations with respect to compliance with discharge standards and long-term stability are high, and demand highly effective solutions on a small area. We also highlight the practical experience of the design, construction and the first years of operation of the Pöhla wetland, which represents a good example. Finally, we discuss some of the pitfalls and potential problems including some realistic cost estimates which are often hidden behind the general, over-optimistic statement of "maintenance-free, zero-cost" passive water treatment systems.

N. Marmiroli et al. (eds.), Advanced Science and Technology for Biological Decontamination of Sites Affected by Chemical and Radiological Nuclear Agents, 177–195.

Keywords: constructed wetlands, passive biological water treatment, adsorbents, uranium, radium, arsenic, iron, manganese

1. Introduction

During the period of the cold war, the Soviet, and later the Soviet-German company WISMUT became the third largest world producer of natural uranium, supplying the Soviet nuclear programme with approximately 230,000 tons of uranium. Following the fall of the Berlin Wall in 1989, WISMUT's operations, which had been losing viability, stopped abruptly, and it was transformed into a company owned by the German Ministry of Economy, specialising in the clean-up and remediation of mining and milling sites (Mager 1996). The German government assigned a budget of around €6.5bn to the WISMUT project. These funds were intended not only for technical tasks, but also to allay the socio-economic consequences of the abrupt end of mining in Saxony and Thuringia, which have faced serious economic challenges as a result of German re-unification. The presence of radioactive contamination over large areas may have contributed to the attention the WISMUT project has attracted since the early 1990's. Since this time, WISMUT has become a key international reference and benchmark for mine closure, rehabilitation and regional redevelopment.

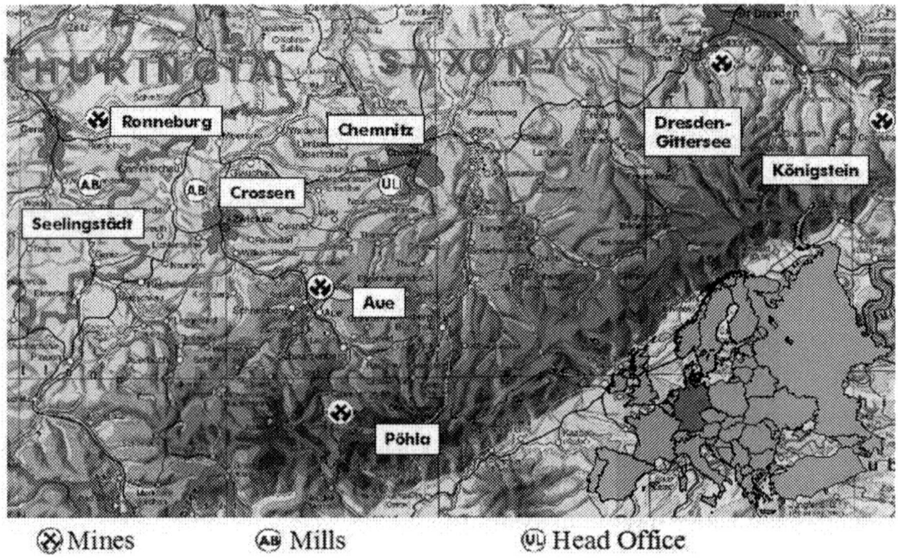

Figure 1. WISMUT mine and mill sites (Inset in the lower left corner shows the location of the WISMUT Project in Europe).

The environmental legacy tackled by the WISMUT project included:

- 3700ha of mining liabilities
- Two large ore processing plants (Seelingstädt, Crossen)
- 311Mm³ waste rock dumps, partly radioactive
- 600ha tailings management (10 tailings ponds)
- Four underground mines (Pöhla, Schlema, Königstein, Gittersee) and an open pit mine connected to underground operations (Ronneburg)
- 31Mm³ per year mine water discharge

The major tasks of the WISMUT project are:

- flooding and geotechnical stabilisation of underground mines, particularly those close to the surface, and those prone to subsidence
- closure and long-term stabilisation of tailings ponds
- *in situ* remediation (including covering) or relocation of waste rock piles
- demolition of production plants, clean-up and redevelopment of the affected property

Flow rates and contaminant concentrations in the effluents of WISMUT sites lie in the following ranges:

- Flow rates
  - Seepage: 1 - 10m³/h, with strong fluctuations of up to 100m³/h
  - Mine effluent: 10.- 1,200m³/h, depending on the catchment area of the mine
  - Free water removal from tailings ponds: about 200m³/h
- Contaminant concentrations
  - Uranium, arsenic: a few mg/l
  - Radium: a few Bq/l
  - Sulphate: 100 - 1000mg/l
  - Nitrate: some 10mg/l
  - Iron, manganese: about 10 mg/l

Water treatment represents one of the largest single cost items of the entire project, being involved in at least the first three tasks listed above. The importance of water treatment derives mainly from the long period of time over which it is required at the various sites, the high flow rates at some sites and, partially, from the complex chemistry of the effluents for which off-the-shelf solutions did not exist (Kießig and Kunze 1996). Apart from consumables, the costs reflect mainly the employment of staff in the water treatment plants.

The effluents arising from the closure of mining and milling sites can be grouped into three classes according when they are generated:

- Effluents from flooded mines are characterised by a more or less constant flow rate but a characteristic decrease of the contaminant load over time, resulting from the initial flush at the time of flooding and a prolonged tail due to the dilution of the mine water body by infiltration water inflow (which is partly overlain by dissolution, adsorption and precipitation processes within the mine)
- Seepage from waste dumps and tailings ponds, whose flow rate is determined by the cover system and, during consolidation, by the hydraulic processes within the tailings body. Likewise, the contaminant concentration depends on chemical reactions occurring within the waste rock and/or the pore water quality of the tailings. In waste rock seepage, both flow rate and chemical characteristics can vary rapidly over time, especially after strong precipitation events.
- Free water removed from tailings ponds: WISMUT has adopted a dry remediation approach for tailings ponds, which requires the initial removal of thc frcc water from the pond, its treatment and finally its discharge into the environment. These waters have high concentrations of common contaminants, although they decrease over time as the pond shrinks and its contents are diluted by precipitation. Rapid removal of the free water is desirable so as to allow an interim cover to be placed over the tailings pond. But this requires a water treatment plant of sufficient capacity, and the treatment technology must be flexible enough to cope with the changing chemistry of the both the free water and, over the long-term, with any dam seepage (if this also needs treatment).

Simplistically, conventional water treatment in mine closure projects becomes less and less efficient over time, because ever-decreasing loads are removed by a technology, which has a cost structure made up of a high proportion of fixed costs (staff, energy), while variable costs decrease only marginally.

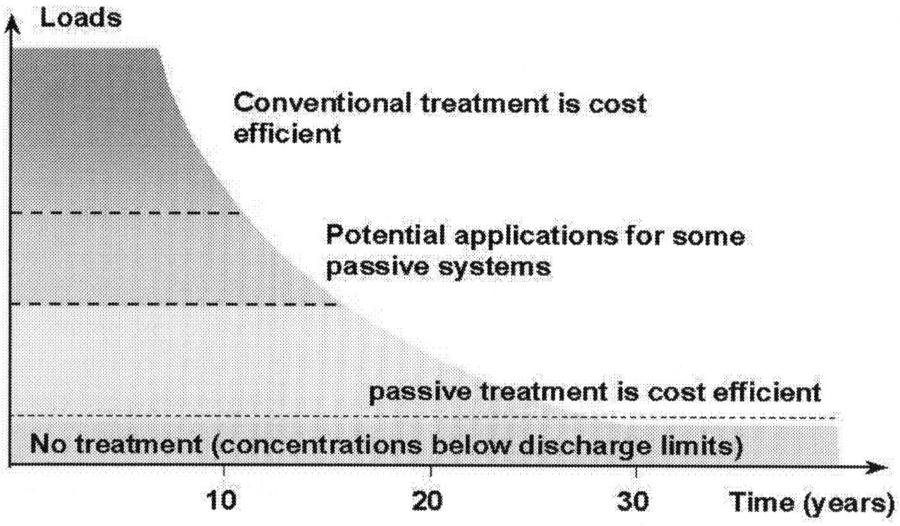

Figure 2. Crossover from conventional to passive water treatment vs. time.

The long-term water treatment strategy must take into account the evolution of both water quantity and quality over time. A strong incentive for passive water treatment systems, given the high overall costs of water treatment in the closure and post-closure phases, is the reduction of long-term costs by the introduction of low-cost, self-sustaining systems requiring only a minimum input of staff and consumables.

The remainder of this paper describes our experience from a particular site in Germany, with its own peculiarities and regulatory framework. The approach we have taken and the conclusions drawn may not be fully applicable to other countries/sites where different conditions may apply. A further important issue is the presence of radioactive contamination, which leads to increased public attention and, thus, possibly to the enforcement of stricter standards than are applied to conventionally contaminated effluents.

2. Conventional and passive-biological water treatment at the Pöhla mine site

2.1. CONVENTIONAL WATER TREATMENT PLANT

The Pöhla mine was the first to be flooded within the WISMUT project, and a water treatment plant was erected and put into operation in 1995. The neutral mine effluent, which is rich in bicarbonate (about 1g/l), has an average flow rate of 17m³/h and has undergone a marked evolution over time since flooding of the mine was completed.

Table 1. Main components of the mine effluent.

Component	Unit	Limit for discharge	1995	1997	1998	2003	2005
Fe	mg/L	2	5	4	4	4	4
Mn	mg/L	2	3.7	1.6	1.1	0.5	0.4
As	mg/L	0.1	0.5	2.0	2.2	2.2	2.9
U	mg/L	0.2	1.8	0.2	0.2	<0.1	<0.1
Ra-226	Bq/L	0.3	1.1	3.9	4.5	4.3	4.3

The treatment technology was based on a selective precipitation/-flocculation process specifically developed by WISMUT in collaboration with leading German research institutions:

- Uranium was adsorbed to GOPUR, a reactive polymeric flocculant
- Radium was co-precipitated as radium/barium sulphate by the addition of barium chloride
- Arsenic was removed using $FeCl_3$
- Iron and manganese were oxidised by aeration
- The resulting residues were de-watered, filled into drums and disposed of in dry parts of the Pöhla mine

The unit cost of this technology, including disposal, was approximately €4/m³. It is estimated that treatment will have to continue until at least 2020. This has prompted our search for alternative, less expensive techniques which take into account expected changes in mine water chemistry, as well as the decreasing load of contaminants over time. Soon after commissioning the conventional treatment plant, it became clear that a passive or semi-passive biological treatment system would be needed for the long-term. Therefore, WISMUT adopted a two-step approach, in which:

- first, a pilot wetland was designed and operated over two years in order to gain experience with biological systems under the prevailing climatic conditions, to optimize the system and to create a basis of trust with the community and the regulatory bodies that a passive system would be a secure alternative to the proven techniques of conventional water treatment.
- second, a full-scale constructed wetland would be built, with its design and operating parameters based on the results from the pilot trial.

A further advantage of installing a pilot wetland was the provision of an experimental basis for passive treatment systems which could be used at other sites within the WISMUT project.

2.2. THE PÖHLA PILOT WETLAND

The pilot and experimentation plant at the Pöhla mine site was built in a concrete pond (part of the former mine operation), with a footprint of $475m^2$ and a total volume of $830m^3$. After the installations were fitted, the usable volume was $415m^3$. The installations divided the concrete pond into seven treatment basins (cells) which served various purposes:

- a small pond to trap coarse matter
- a sedimentation pond
- a gravel-filled pond for aerobic-anaerobic conditions
- a gravel-filled pond for anaerobic-aerobic conditions which can also be vegetated with plants
- a polishing/oxidation pond with plants
- two small ponds which can be filled with reactive material (adsorbents)

Before reaching the wetland, the mine water was fed through an aeration cascade (see Figure 5), bridging the geodetic height difference of around 20m between the mine adit and the pond. The pilot wetland was put into operation in 1998, with an average flow rate of $2m^3/h$ treated mine effluent. Floating mats and similar installations were used to assist the sedimentation of iron precipitates to which arsenic is mainly bound (Figure 3). These have proved to significantly improve the effectiveness of iron and arsenic removal by minimizing the pond size, an essential feature given the limited space available at most sites. Figure 3 shows the principal elements and their layout, while Figure 4 shows a global view of the constructed wetland.

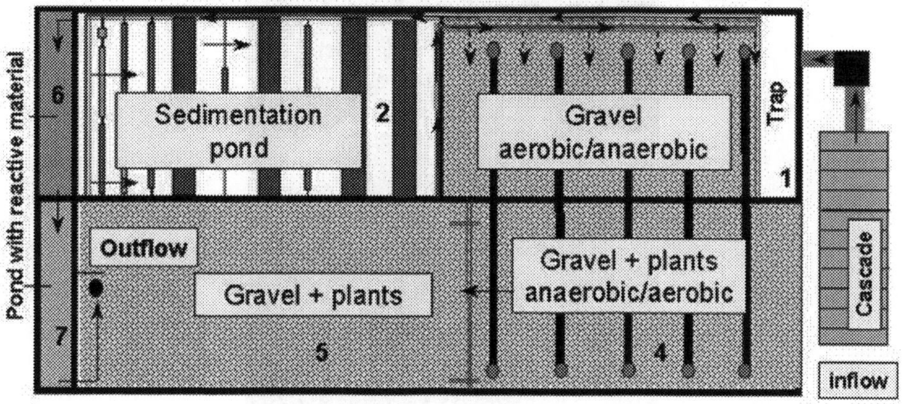

Figure 3. Layout of the design elements (cells), with the aeration cascade on the right-hand-side.

Figure 4. View of the pilot wetland, the aeration cascade is visible in the left lower corner. In the foreground, there are plastic ponds used for Radium sorption tests with algae.

Other special features used in the pilot wetland are floating mats and similar installations to assist the sedimentation of iron precipitates to which arsenic is mainly bound. They can be seen in Figure 6. They have proved to significantly increase the effectiveness of iron and arsenic removal by keeping the pond size to a minimum. This is essential under the conditions already described in the Introduction, namely limited space available at most of the sites.

Figure 5. Aeration cascade between the mine adit and the inflow to the pilot wetland. The dark stain from iron precipitates is clearly visible.

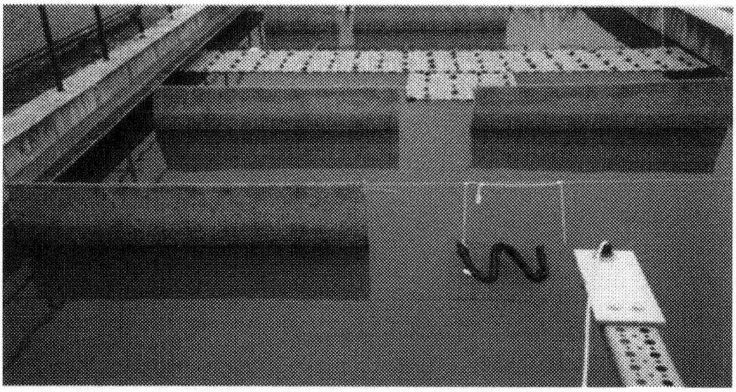

Figure 6. Floating mats in the pilot wetland (so-called "Aquamats").

A parallel R&D project ("BioRobust") investigated the robustness and resilience of passive water treatment systems under the site-specific conditions at WISMUT (Kunze et al. 2002). This study showed that although small wetlands (which must concentrate their functionality on a very small area) can operate reliably under a wide range of external conditions and achieve permissible discharge standards, their performance also tends to fluctuate if temperature, flow rate or the chemistry of the mine water at the inflow varies beyond certain limits. This has proved to be the case at the Pöhla mine site. Peaks of arsenic concentration from the mine adit were observed which correlated with flow rate fluctuations due to hydraulic conditions in the mine, and these have led to the exceeding of permitted limits of the discharged water. As a result, counter-measures need to be developed to guarantee compliance with discharge limits, since this is an important requirement for the granting of an operating permit and for consent from the authorities to phase out the conventional water treatment plant. Adsorbents placed in the last two small sections of the pilot wetland (ponds 6 and 7 in Figure 3) are essential to retain residual concentrations of arsenic and radium which were not fully removed by the wetland. The following materials have been successfully used in the pilot wetland to smooth fluctuations in performance:

- granulated barium sulphate in a chemically inert matrix of alumino-silicate binder (trade-name Hedulat), developed and patented by WISMUT (Hermann et al. 2001, see also Kunze et al. 2002a)
- granulated ferric hydroxide (trade-name FerroSorp), see Kasting (2005)

Apart from the sorption capacity and cost, an important property for these materials was their ability to retain the adsorbed contaminants during and after solidification by a hydraulic binder, as all wastes (including spent adsorbents) from the wetland have to be disposed of remotely. The central disposal site for the Schlema mine water treatment plant is some 15km from the Pöhla site (see Figure 1), and this plant is expected to operate for at least the next 20 years, i.e., beyond the treatment time for Pöhla mine water. Thus all wastes generated from the Pöhla pilot (and, later, full-scale) wetland will be disposed of at Schlema.

The Pöhla pilot served as a basis for the design of the full scale wetland, in particular with respect to:

- removal rate per unit area in the various treatment cell types
- hydraulic conductivity of the various cell types
- function of the floating mats and/or floating biomass with extensive roots (see Smith and Kalin 2000)
- measurement of performance fluctuations and their correlation to external factors
- dimensioning of polishing filters (adsorbents)
- sampling cycles for drafting the monitoring plan

Another important function of the pilot wetland was its demonstration of performance to regulators, setting the basis for the negotiation of achievable and permissible discharge concentrations.

2.3. MACROPHYTIC ALGAE AS HYPER-ACCUMULATORS FOR RA

While uranium contamination has nearly vanished from the effluent due to the reducing conditions present in the mine, radium has increasingly become a concern. A peculiarity of radium is that although it occurs in mine water in significant activity concentrations (some Bq/l), this corresponds to a very small mass concentration (1Bq of Ra-226 corresponds to 270pg). Macrophytic algae from the Characeae family (Figure 7 and 8) have been shown to be hyper-accumulators of radium (Kalin et al. 2002). Although the mechanisms for Ra incorporation are not understood in every detail, it is thought that Ra ions are loaded into the calcium-rich lattice and/or form of $BaSO_4$ substitute crystals in the algal tissue. A more detailed account of recent research in this area is given by Kalin et al. (2002a).

2.4. THE FULL-SCALE PÖHLA WETLAND FOR MINE WATER TREATMENT

In 2003, the full-scale wetland was built and its operation began at the Pöhla site. Figure 9 illustrates that it consists of two separate, independent lines, each with three sequential ponds and polishing filters containing

Figure 7. Macrophytic algae *Characeae vulgaris* which are used as hyper-accumulators for Radium from water.

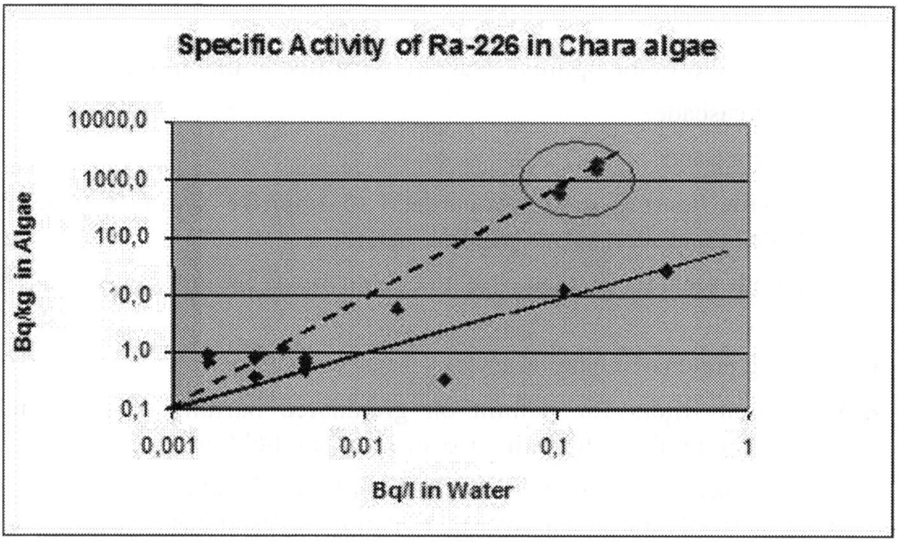

Figure 8. Specific activity of Radium in ordinary plants and algae (solid line) compared to Characeae algae (dotted line), showing roughly a linear relationship with the Radium activity concentration in the water.

reactive material. The double line design is required to achieve adequate redundancy. The ponds have a total area of approximately 3100m^2 and a total pond volume of 2200m^3, and the installation treats an average of 17m^3/h of mine effluent, with a maximum design capacity of 20m^3/h.

Figure 9. Aerial view of the full scale wetland at the Pöhla site. The double-line structure can be clearly seen, with one line in the upper part of the photograph and the second line in the lower. In the left lower corner, the aeration cascade is visible, followed by the rectangular settlement pond (concrete structure).

The wetland follows the optimal structure identified in the pilot experiment:

- Aeration cascade
- Settlement pond
- Pond with floating mats (Aquamats) to improve sedimentation of suspended precipitates (two parallel lines)
- Pond with Characeae algae for the removal of radium (two parallel lines)
- Polishing pond (two parallel lines)
- Adsorption filters for radium on granulated barium sulphate and arsenic on granulated ferric hydroxide (two parallel lines)

The monitoring programme reflects the still novel status of the wetland and requires substantial manpower and laboratory resources. It was agreed with the regulators (both radiation protection and water resources authorities) and includes the following sampling points and frequencies:

Parameter	Constructed Wetland		Outflow			
	Inflow	Outflow	Pond 1A/B	Pond 2A/B	Pond 4A/B	Filter 6A/B
As	w	w	m	m	m	m
^{226}Ra	w	w	m	m	m	m
Fe	w	w	m	m	m	
U	w	w	m	m	m	
Mn	w	w	m	m	m	
Cl	w	w	m	m	m	
SO_4	w	w	m	m	m	
NO_3	w	w	m	m	m	
TSS	w	w	m	m	m	
P		m				
Cr	q	q	q	q	q	
Zn	q	q	q	q	q	
Cd	q	q	q	q	q	
Pb	q	q	q	q	q	
Hg	q	q	q	q	q	
Ni	q	q	q	q	q	
Se	q	q	q	q	q	
Cu	q	q	q	q	q	
Ba	q	q	q	q	q	
Na	q	q	q	q	q	
K	q	q	q	q	q	
Mg	q	q	q	q	q	
Ca	q	q	q	q	q	
Al	q	q	q	q	q	
CO_3 / HCO_3	q	q	q	q	q	
PO_4	q	q	q	q	q	
TWH[1)]	q	q	q	q	q	
TDS	q	q	q	q	q	
N / NO_2 / NH_4	q	m	q	q	q	
COD	q	m	q	q	q	
BOD_5		m				
radionuclides	q	q	q	q	q	
microbiology	q	q	y	y	q	

w weekly
m monthly
q quarterly
y yearly
1) Total water hardness

The current analytical programme generates about 3000 samples per year. Over time and with increasing statistical material, the extent of the analytical programme is expected to decrease.

3. Economic and management issues

3.1. COST STRUCTURE

In order to assess the cost advantages of the constructed wetland over the existing conventional water treatment plant, the critical measure is the pay-back period for the one-off capital costs of construction. Thus, the construction cost of about €700,000 has to be factored into the difference in operating costs between the conventional water treatment plant and the wetland.

- Operating cost of conventional plant - €4/m³
- Operating cost of new wetland
    - first years after construction: approximately €2/m³ (where the limited activity of algae and aquamats must be compensated for by a higher loading of reactive filter materials)
    - long-term, steady operation: approximately €1 - 1.50/m³

Using the higher operating cost of €2/m³ and an average flow rate of 17 m³/h, the annual cost savings are about €300,000, resulting in the repayment of the capital cost within two to three years. However, a single figure for the operating costs is misleading, as it obscures the difference between fixed and variable (i.e., proportional to the flow rate or contaminant load) costs. It is therefore worthwhile to look at these components separately:

Fixed costs:

- In the first years of operation, operating personnel are required to a level of 75 h/day, seven days per week for monitoring (note that chemical and radionuclide analyses are not included in this costing), visual inspection and minor repair works. These expenses will be reduced over time when more statistical data become available from the wetland operation. Labour costs are estimated at about €30/h.

- Supervision of the operation by engineering staff (including reporting, review and evaluation of sampling data, and administration): 14 h/week at €50/h.
- General maintenance tasks, including adjustment of flow meters and measuring equipment, cleaning ponds, mowing grass and winter service (some of which are outsourced).
- Miscellaneous items, including telecommunication fees, vehicle costs, insurance and sundry other expenses not related to the flow rate or contaminant load treated by the wetland.

Variable costs:

- Replacement of reactive filter material, including the disposal, handling, transport and solidification of spent filter material. Importantly, some of the spent filter granulate (especially the Ra adsorbent) must be transported according to ADR class 7 regulations ("Radioactive material", see Kunze, 2003) which adds considerably to the cost due to requirements for special vehicles and safety precautions. The removal of spent granulates from the filter tanks requires special vacuum equipment, a necessity that was not taken into account in the design phase but in practice was complicated. Although this is a major component of the cost, it is expected to decrease significantly over time because the physical and biological components of the wetland are expected to be sufficient to achieve the discharge limits without any reactive filters.
- Replacement of plants, floating mats etc., including removing dysfunctional algae and growing/planting new ones.

Some time is needed for the constructed wetland to operate smoothly and at the intended low cost level. In particular, the algae and floating aquamats need several years to reach full performance, so that the filter adsorbents are needed during this period. The frequent replacement of these filters adds considerably to the cost. Stand-by of the conventional treatment plant, which is required by the regulators to cope with any unexpected malfunctioning is another cost factor which makes the wetland more expensive in the early years of its operation. The cost breakdown is summarized in Table 2, which also indicates those components which can be significantly reduced over time.

Table 2. Summary of operating cost breakdown.

Cost item	Value (T€ p.a.)	Cost per m^3 treated (at an average flow rate of 17 m^3/h)	Potential of cost savings over time (see Figure 10)
Fixed			
Workers	41	0.28	yes (0.20)
Engineers	36	0.24	yes (0.12)
Maintenance	10	0.07	no
Miscellaneous	26	0.17	no
Subtotal fixed costs	*113*	*0.76*	*0.56*
Variable			
Replacement and disposal of spent filter material	-	1.20	yes (0.60)
Replacement of plants, floating mats etc.	-	0.02	no
Subtotal variable costs		*1.22*	*0.62*
Total		**1.98**	**1.18**

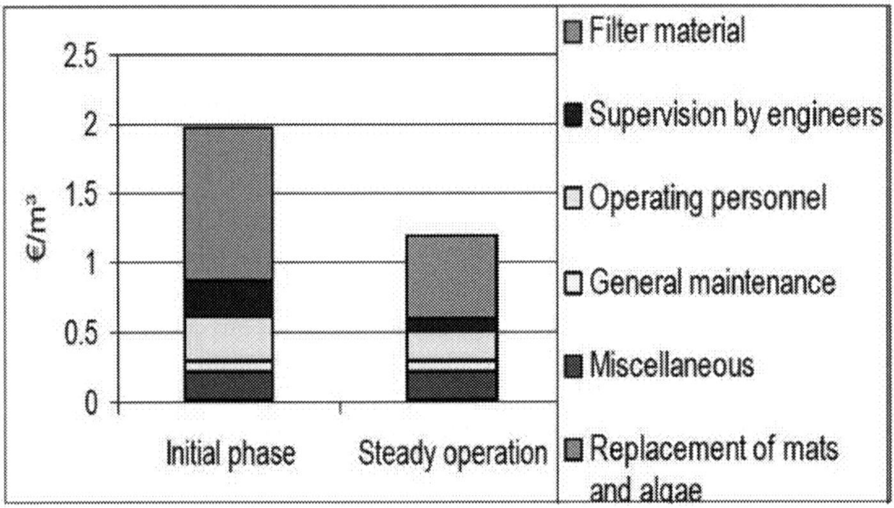

Figure 10. Comparison of the breakdown of operating costs in the initial phase and as expected over the long-term.

3.2. WASTE DISPOSAL ISSUES

The question which often arises in the context of passive biological water treatment systems is that of waste and waste disposal. The following wastes have to be handled and disposed of in the Pöhla wetland:

- Ferric hydroxide sludge from aeration cascade
- Spent filter material
- Defective floating mats
- Algal debris

The first three wastes are solidified in the Schlema Water Treatment Plant. The costs of solidification and subsequent disposal in a landfill cell over a covered uranium waste rock pile are negligible (approximately €0.01/m^3). However, as mentioned earlier, the technical complications of handling the spent reactive filter material may be considerable and lead to unexpected costs. In particular, the "sucking" of material out of the filter tanks was non-trivial, due to the rheological properties of the filter granules. Moreover, transport of the radioactive material has to be carried out according to Class 7 ADR regulations, which set certain requirements for the vehicles used, especially as the filter granulate is wet. The handling and disposal of the hydroxide sludge and the floating mats have so far been straightforward. The amount of Characeae debris is negligible and also does not constitute a major cost component. The algae mineralize and form an inorganic layer at the bottom of the pond. Estimates show that the depth of accumulated algal debris will reach 15cm over a 10 year period. An issue yet to be resolved will be how the mineralized algal sludge can be removed without damaging the algae. However, it is likely that there will be no need to remove the algal sludge, as the expected operating time will not exceed 15 years.

3.3. OTHER IMPORTANT OPERATIONAL ISSUES

In the operation of the Pöhla wetland, a number of other issues have arisen which are briefly addressed here:

- Responsibility for mine water quality and quantity: From time to time, the quality and flow rate of the mine water fluctuate significantly, and this may cause overloading of the entire system and subsequent non-compliance with permitted discharge concentrations, which can be only be buffered by the reactive filter materials. In this respect, contractual questions must be resolved between the operator of the wetland (WISUTEC) and the owner of the mine water (WISMUT) as to who is responsible for treatment and discharge.

- Supply of algae: The hyper-accumulating algae are in insufficient supply, as they appear on the Red List of endangered species. Therefore, the operator must take measures to grow the algae in the required quantity and quality at its own cost and risk.
- The reactive filters suffer from hydraulic blockage due to calcite crystal staining of the filter granules. Additional equipment (compressors for pressurized air with which the filters can be periodically back-flushed) had to be installed. Not only does this lead to higher costs due to depreciation of the equipment and more manpower, but also is inconsistent with the overall concept of a passive, maintenance-free system.
- The calcite staining of the filter granules also leads to a significant decrease in the sorption capacity of the reactive filter material. This, in turn, leads to more frequent replacement cycles and higher costs.
- The problems with the filter materials have also led to a more frequent reporting requirement by the regulators than was initially planned.
- Organisational issues such as the information chain in case of an emergency or technical problems, responsibility for work and safety instructions, must be clearly resolved between operator and the owner of the wetland.

4. Conclusions

The treatment of mine effluents over the long-term using a passive wetland system is possible, and the cost savings are considerable. However, the notion of a maintenance-free, "zero-cost" system is not tenable in the light of our experience. Certain costs are unavoidable (maintenance and monitoring requirements), but others relate to technical measures required to guarantee the performance of the system and its compliance with the strict permitted discharge limits. One of the major outcomes of our work is the identification of numerous unexpected complications with the hydraulic and sorption properties of the filter materials, and the technical solutions for their replacement. Such seemingly trivial issues can lead to additional technical efforts which would be negligible in the case of a conventional water treatment system, but are disruptive in a system which is based around the concept of passiveness and minimum human intervention. The cost advantage of constructed wetlands may, at least in its early years, be small or non-existent, since the stand-by of the conventional treatment plant is required until the passive system reaches full performance.

References

Kießig, G., and Kunze, C., 1996, Wasserbehandlung und Rückstandsentsorgung (Water Treatment and Waste Disposal, in German), Geowissenschaften, Vol. 14, Issue 11, November 1996, p. 481.

Hermann, E., Dullies, F., Griebel, I., and Kießig, G., 2001, Filter material and process for the removal of heavy metals, arsenic, uranium, and radium from contaminated waters, Patents Pending: DE 101 16 951, DE 101 16 953 (2001).

Kunze, C., Glombitza, F., Gerth, A., Kiessig, G., and Küchler, A., 2002, Long-term stability and resilience of passive mine water treatment facilities: A joint experimental and simulation approach. In: Merkel, B.J., Planer-Friedrich, B., Wolkersdorfer, C. (Eds.): Uranium in the aquatic environment. Proceedings of the International Conference Uranium Mining and Hydrogeology III and the International Mine Water Association Symposium, Freiberg, Germany, 15-21, September 2002, Springer Verlag, Berlin, Heidelberg, New York, p. 597.

Kunze, C., Hermann, E., Griebel, I., Kiessig, G., Dullies, F., and Schreiter, M., 2002a, Entwicklung und Praxiseinsatz eines hocheffizienten selektiven Filters für Radium (Development and practical application of a highly efficient selective Radium filter, In German). GWF Wasser Abwasser, Vol. 143, Issue 7-8, 2002, p. 572.

Kunze, C., 2003, Handling, treatment and disposal of naturally occuring radioactive wastes (NORM) from mining and water treatment. In: Hebenstreit, C., Kudelko, J., and Kulczycka, J. (Eds.): Mine Waste Management – Best Available Techniques, Krakow 2003, p. 186.

Kalin, M., Kießig, G., and Küchler, A., 2002, Ecological water treatment processes for underground uranium mine water: Progress after three years of operating a constructed wetland. In: Merkel, B.J., Planer-Friedrich, B., and Wolkersdorfer, C. (Eds.): Uranium in the aquatic environment. Proceedings of the International Conference Uranium Mining and Hydrogeology III and the International Mine Water Association Symposium, Freiberg, Germany, 15-21, September 2002, Springer Verlag, Berlin, Heidelberg, New York, p. 587.

Kasting, U., 2005, Reinigungsleistung von Filtersubstraten zur Behandlung der Abflüsse stark befahrener Straßen (Purification performance of filter substrates to treat the runoff from roads with heavy traffic, In German), Technische Akademie Hannover, Newsletter 02/05 (2005).

Mager, D., 1996, Das Sanierungsprojekt WISMUT (The WISMUT Remediation Project. In German), Geowissenschaften, Vol. 14, Issue 11, November 1996, p. 443.

Kalin, M., Smith, M. P. and Wittrup, M. B., 2002a, Ecosystem restoration Incorporating Minerotrophic Ecology and Stoneworts that Accumulate 226 Ra, In: Merkel, B.J., Planer-Friedrich, B., and Wolkersdorfer, C. (Eds.): Uranium in the aquatic environment. Proceedings of the International Conference Uranium Mining and Hydrogeology III and the International Mine Water Association Symposium, Freiberg, Germany, 15-21, September 2002, Springer Verlag, Berlin, Heidelberg, New York, p. 495.

Smith, M., and Kalin, M., 2000, Floating Wetland Covers In Biological Wastewater Treatment Systems. Quebec 2000 Millennium Wetland Event, Quebec City, Canada, August 6-12, 2000.

LESSONS FROM THE WASTEWATER FIELD

DAVID L. RUSSELL
President, Global Environmental Operations, Inc. 4642 Warrior Trail, SW, Lilburn, GA, 30047, Phone:770-923-4408: Fax: 770-381-8004, e-mail: dlr@mindspring.com

*To whom correspondence should be addressed.

Abstract- The wastewater treatment field has over 100 years of dealing with chemicals in groundwater, and in consistently removing over 95% of the contaminants. This paper will discuss some of the basic principles involved and ways in which wastewater lessons may be applied to remediation technology.

Keywords: Remediation, wastewater, lessons

1. Introduction

In every developing field of knowledge, there is a tendency to re-invent the wheel because the field is new and the particular application has never been handled in that manner before. A good example of this is the field of groundwater flow. The equations, even for layered structures are essentially heat flow equations, and the applicability is there, but because groundwater geologists were not familiar with the literature of thermal flow and chemical engineering, many of the models and the works of the chemical engineers went unused while much effort and time was spent in re-discovering material transport and the flow equations which were first solved for thermodynamic systems.

I believe that many of the same problems exist in the remediation field. Most remediation specialists don't consider what the wastewater treatment people have been doing with bacterial systems for the past 100 years. Not all the work in wastewater treatment may be applicable to remediation, but it does not deserved to be dismissed out of hand either. In many cases, what is happening involves water, oxygen, nutrients, a substrate, and the same soil organisms which are found in wastewater. In fact, one of the ways of

N. Marmiroli et al. (eds.), Advanced Science and Technology for Biological Decontamination of Sites Affected by Chemical and Radiological Nuclear Agents, 197–208.

beginning a wastewater culture for a specific compound is to start with soil which has been contaminated with the compound. This was one of the ways in which the wastewater people found that they could and did develop organisms which could successfully degrade pentachlorophenol, and many biphenyls, compounds normally hard to acclimatize bacteria to. In fact, I'm aware of some research which is being conducted currently using soil cultures, on heavy petroleum fractions, and growing the acclimatized bacteria in liquid culture and then reapplying it to the soil for enhanced degradation.

2. First principles:

2.1. RUSSELL'S LAW

There are a number of principles which the wastewater professional are aware of, but which the remediation professionals need to consider. The first of these is intent. What I've called Russell's First Rule of Wastewater Treatment is: "Given any combination of organisms, nutrients, substrate, and oxygen, the organisms will do precisely as they please for their own best interests. It is our task to make our interests and their interests coincide." One of the corollaries to this rule is, "Bacteria are like children with candy. When the candy is available they will postpone dinner until the candy is all gone." Bacteria will most often go for the simple sugars and starches first because they are easiest to digest and every self respecting bacterium has enzymes for this digestion already, and they don't need to create anything special. It is not until that food runs out that your bacterial "children" will eat their broccoli or develop the enzymes needed to hydrolyze more complex organic substances.

2.2. THE BALANCED DIET

Wastewater professionals usually measure aerobic bacterial degradation by measuring the oxygen consumption. The standard measure for this is a not very well understood but widely used measure called Biochemical Oxygen Demand. The test measures the amount of oxygen depleted from nutrient rich water based 5 days. The test is widely used, but has many limitations including the inability to be accurate at values of under 3 mg/l. Ideally the test measures the bacterial conversion of carbon compounds, but can also measure nitrification unless an inhibitor is used. So, if you have a compound which has a 5 day BOD of 600 mg/l, and you spill it on the soil, every liter of material will want to consume about 600 mg of oxygen. Since the oxygen has to come from someplace other than the water, because water has only about 16 mg/l maximum of dissolved oxygen at optimum conditions, it is safe to assume that the soil chemistry will undergo a change as the oxygen is depleted. The ORP of the spilled liquid and the soils will become strongly negative as anaerobic degradation takes place. If the anaerobic digestion is fast enough, the soil/ liquid may form volatile fatty acids and develop an acid pH and essentially stop further biological activity until other organisms can come in an develop the enzymes to degrade the VFA or further convert them into methane (anaerobically) or aerobically reduce the VFA to CO2 and H2O. If the action is anaerobic, it will consume alkalinity.

The point is that bacterial populations which are the workhorses for our degradation need a balanced diet. If we measure the C:N:P ratio of a healthy bacterial population, we find that it is about 100:5:1 if the carbon demand is expressed as BOD5. If we express the CNP ratio in terms of COD, the numbers become 150:5:1. If we look at recommended stoichometric ratios for biological degradation, we wind up with numbers, when the Carbon is expressed as BOD5 vary between 17-32 BOD5/N, and 150-90 BOD5/P. Much of this work is based upon a biomass of $C_5H_7NO_2$. For complex systems, it is not uncommon to calculate the stoichiometry of the waste stream, and of the bacteria and attempt to balance the nutrients required to a given minimum for the growth[4].

In the process of generating biological growth you will need to watch the alkalinity of the soils, even when localized. In achieving nitrification, ammonia is consumed, but so is about 1.83 moles of Alkalinity as HCO3- for every mole of Ammonia consumed by nitrosomanas and nitrobacter.

2.3. THIRD: GROWTH RATES AND REACTOR KINETICS

The classic expression for measurement of growth rate is the Monod Equation.

$$dx/dt = \mu = \mu \max * S/(Ks+S)$$

Where μ = specific growth rate, μ max = maximum specific growth rate, X = microorganism concentration, S= growth limiting substrate concentration, and Ks= half saturation coefficient for hydrolysis.

Graphically, the determination of μ looks like in Figure 1.

[1] Orohon, Derin, and Artan, Nazik, Modeling of Activated Sludge Systems, Technomic Publishing AG, Missionstrasse 44, CH 4055, Basel, Switzerland

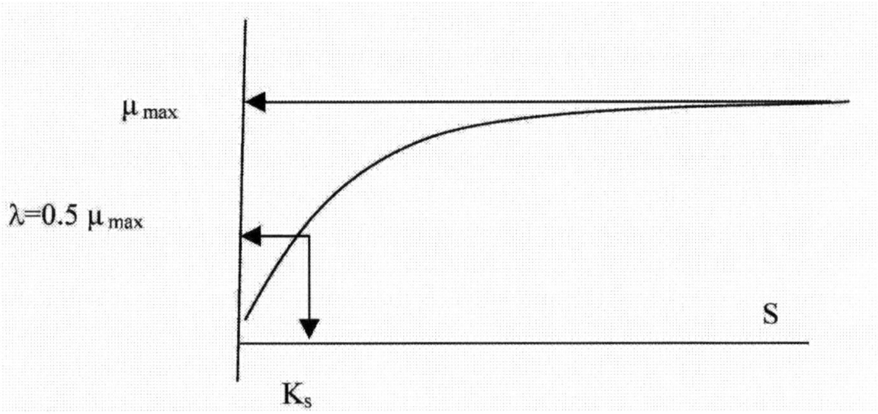

Figure 1. Graph of Monod Equation.

Where μ is determined from the logarithmic growth rate.

To this was added reactor hydraulics and microbial residence time (sludge age) and a given reactor configuration, resulting in the following Figure 2.[2]

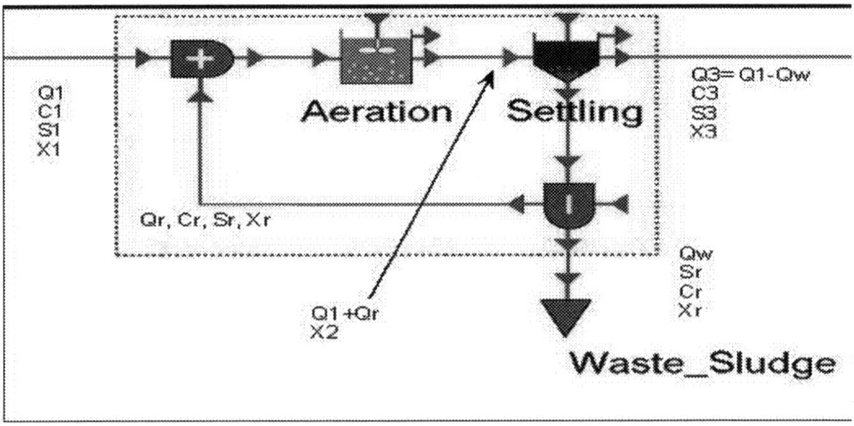

Figure 2.

Given the reactor shown above, the following terms are applicable.

First we will look at some classical definitions and then at a derivation for the Activated Sludge Equations for ASM1

Defining terms first:

Q = Volumetric influent rate – volume/ time

Q_w = Waste sludge volumetric flow rate – volume/ time

Q_3 = Effluent Flow rate

Q_r = Recycle Flow Rate

X_1 = Microorganism Influent Concentration -- mass/ volume-- influent

[2] Russell, DL. Chapter 6, Notes on Wastewater Treatment, published by Am. Institute of Chemical Engineers, NY, NY, to be republished by J. Wiley & Sons, NY, 2006.

X_2 = aeration basin microorganism concentration – mass/ volume
X_3 = Secondary Effluent microorganism concentration – mass/ volume
X_r = Recycle and Wasted Solids Concentration
V_2 = Aeration Basin Volume
r_{BH} = reaction rate for solids also may be written as dX/dt = rate of change of microorganisms concentration in aeration basin – mass/ volume-time

r_s = reaction rate for substrate.

Rate of Bacterial Growth $r_{BH} = \mu X$ where X is the microorganism concentration in mass/ volume

and μ = specific growth rate per unit of time

$$\text{Cell Yield Coefficient} = \text{Yobs} = -\frac{\mathbf{r}_g}{\mathbf{r}_{su}}$$

Where Y obs = observed yield coefficient
and $\mathbf{r}_s$ = substrate utilization rate; $\mathbf{r}_{BH}$ = Cell growth rate

and $\mathbf{r}_{BH} = -Y_{max}\mathbf{r}_s - bX$

where Y_{max} is equal to λ, and
b is the specific maintenance rate, endogenous or decay coefficient in units of Time.

This gives us a sample of a solution for a steady state system.

Now when we look at a biological treatment system, we'll consider a simple system as comprised of a reactor or aeration tank, and a clarifier or solids removal device, as shown below.

Running a balance around the system we get:

$$Q_1X_1 + VX_2r_2 = Q_3X_3 + Q_wX_r$$

if X_1 is relatively small with respect to X_2 and we assume steady state operations, then the equation becomes:

$$\mu = r_2 = \frac{Q_3X_3 + Q_wX_r}{V_2X_2}$$

For a bioreactor Mean Cell Residence time = SLUDGE AGE = θ_c = Solids Mass/Change in solids mass = $X/(\delta X/\delta t)$ or

$$\theta_c = 1/\mu = \frac{1/r_2 V_2 X_2}{Q_3 X_3 - Q_w X_r}$$

One measure of activated sludge systems is the mean cell residence time or sludge age. The different types of systems and much of US terminology is involved with Sludge Age.

Again, at steady state conditions, and making a substitution from above we get:

$$\text{Yobs} = \frac{\theta X}{\theta_c (S_o - S)} \text{ and } \frac{\text{Ymax}}{1 + b\theta_c} = \frac{\theta X}{\theta_c (S_o - S)}$$

$$\text{Specific Utilization Rate} = U = \frac{S_o - S}{\theta X}$$

With one other critical substitution of Efficiency $E = (S_o - S)/S_o * 100$

We get $U = F/M * E * 10^{-2}$

Where F/ M is the Food to Microorganism Ratio or

$$F/M = S_o/\theta X$$

The F/M ratio is one of the key parameters in designing an aerobic treatment system by conventional means in the US. This is also called loading rate.

The point is that the same population dynamics are applicable to remediation systems. The principal difference is that in a soil/water system, one has essentially the growth rate of the bacteria as a limiting condition. This is also akin to another type of system known as the Sequencing Batch Reactor or SBR.

The SBR has achieved wide popularity in the US principally because it is cheap and simple to operate. It has no clarifier and only an aeration tank which also serves as a settling tank. A simple SBR Cycle looks like the following Figure 3.

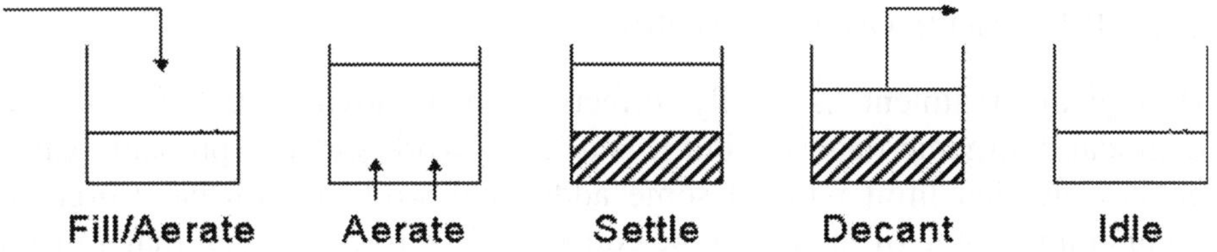

Figure 3.

Without the processes of settling and decanting (and wasting sludge), this is precisely the same process as treating a contaminant in a soil site. Allowances have to be made for water balance, but the process is the same.

But what about the feed? The feed, municipal sewage, or industrial waste often contains both a source of nutrients to permit it's own self growth, and a source of bacteria. Industrial waste, however, contains none of these and frequently has to be adjusted for C:N:P balance and removal of specific toxics.

2.4. FOURTH: THE MODELS

The IWA (International Water Association), formerly known as the IWQA, has had several task forces working on model development for various types of processes. I believe that these reactor models have a good potential application for remedial treatment. The subject of the models is extremely complex and too involved for this discussion, as it is a Master's Level course in Environmental Engineering. However, let me indicate that there are several types of models which may have some application to the bioremediation field. The principal models are

ASM1- A basic model dealing with nitrification and denitrification in wastewater,

ASM2 & 2d, for phosphorous removal

ASM3 A variant of ASM1 with a different, decoupled hydrolysis mechanism

ADM1 The Anaerobic Digester Model

The models may have some utilization in the field of bioremediation.

There are also a lot of free models available on the Internet. One such is the model from Clemson University which was developed by Grady Harmon. It is an elemental model of the wastewater treatment process using ASM1, and it is COD based, but it is free.

2.5. FIFTH: MORE OBSERVATIONS

Biological treatment is easily effective in removing 95-98% of the degradable organic material, but it is a lot of work and you probably can't go beyond that limit without some additional work. If you have organic compounds, some of them may be considered "refractory", but that's just a short way of saying that they will require more work and a longer time to degrade. Don't forget that most biological waste treatment processes are conducted in a realm of 6-24 hours retention time in contact with bacteria in a liquid medium.

Depending upon how you measure your contaminants, you will not get all the COD from the material because COD contains refractory compounds, and you can get a good indication of how your biosystem is doing by looking at the COD.

Biological treatment systems do not handle shock loads well. That probably won't be a problem because you are dealing with a soil system and not a liquid system, but the principles are the same. It will take longer and substantially more work to degrade higher concentrations of materials in the soils.

2.6. SYSTEM LIMITS

There is an apparent upper limit to the strength of waste a biological system can handle in a liquid system. In some cases the constraint is oxygen transfer. In other cases it is the solids concentration in the mixed liquor and the shear which is necessary to keep it in suspension. Recent work with membranes replacing the clarifiers have indicated that the upper limit to solids is between 2% and 4% total solids, but not all of that is active biomass. The solids retention time in the system is often well over 40 days. The apparent upper limit on waste expressed as BOD is about 600mg/l - 800 mg/l. Stronger wastes can be treated, but they comprise a portion of the total flow and in effect are diluted.

Toxic and biologically resistant materials will require special consideration for their treatment. You will need to adjust the nutrient stream to accommodate the bacteria in the system and aid in the hydrolysis of the compounds or even wash or chelate the toxic metals out of the way. In one waste stream where nitroalcohols were being treated, the system required 42 days of detention in order to provide sufficient dilution and residence time to allow specialized enzymes to develop in the bacterial population.

In other cases, specialized enzymes and genetically engineered bacteria have been used to gain temporary relief of shock and other toxic loads to a treatment system. The application of engineered bacteria may be practical in a limited environment, but experience has shown that the bacteria are substantially more expensive and they tend to loose their "engineered" properties after 4-10 generations, so they have to be continually reapplied in order for the benefit to be maintained. It has been shown that in these types of specialized situations, one is far better off to develop local cultures through treatability studies.

3. Sludge and off streams: Handling residuals and effluents

Biosystems are remarkable systems for what they do. They can remove nitrogen, reduce phosphorous and remove it from wastewater, and destroy any wide variety of organic compounds by turning them into CO_2 and water. However, they cannot create nor destroy matter. If you have toxics, you will need to handle the toxics in a pretreatment system or somehow dilute them down to the level where they can be treated. Metals are a particular problem because they bio accumulate even in bacterial systems.

When we look at biological systems, the problem of re-release is particularly critical. In wastewater treatment Nitrogen control and Phosphorous control have been identified as critical elements in preventing algal blooms downstream from wastewater treatment plants. Part of the problem in designing the wastewater process is control of the re-release of these compounds. Nitrogen can be reduced back to a gas, but Phosphorous has to be treated by precipitation to remove it from the wastewater stream. The same is true for almost any of the heavy toxic metals such as Arsenic, Lead, Copper, Uranium, and Cadmium to name a few. Safe to say, this is also a common problem with phyto-remediation systems.

4. Anaerobic systems

In some cases, phyto-remediation may involve anaerobic processes. In wastewater treatment plants, wastes with measured BOD above 500 are generally a good candidate for anaerobic treatment. In anaerobic processes there are three parts:

Fermentation of the wastes – conversion to acetates

Acetogenesis – conversion to acids, formaldehyde and & Hydrogen, and

Methanogenesis – conversion of formaldehyde, acetates, and acids to CO_2 and Methane

One of the principal challenges to anaerobic treatment is balancing the rates of growth. The acid forming bacteria operate at about 3 times the rate of the methane forming bacteria, and without a balanced microbial population the wastes will turn acidic and all methane production will stop.

In sludge digester startup, this condition is known as a “stuck digester” and can be cured by the slow addition of alkaline buffers to the mix. Strong alkalis can take the mix well out of the sludge range, where all activity stops.

Anaerobic fermentation can occur in the pH range of between 5 and about 9, while the methane bacterial operate in a much narrower range of between 6.5 and about 7.6, with the optimum range at about 7.0.

General formulations for anaerobic decomposition have been provided by Buswell for carbohydrates:

CnHaOb+(n –a/4 –b/2)H2O à (n/2 – a/8+b/4) CO2+(n/2+a/8 –b/4)CH4

Most of the bacterial acids formed are more commonly proprionic and acetic acids. Another researcher, Perry McCarty estimates the following:

Amino and Fatty Acids	A = 0.054F – 0.038 M
Carbohydrates	A = 0.46F – 0.088M
Nutrient Broth	A= 0.076 F – 0.014M

where A= biological solids accumulated
M= ML VSS
F= COD utilized.

The wastes must have a balanced feed, including freedom from high concentrations of salts, and relatively high levels of alkalinity must also be present to counteract the CO_2 generated.

Sludge feed varies between 1% and 3.5% solids being fed to the digester. The quantity varies depending upon the system and the amount of sludge which needs to be wasted per day to reduce the amount of solids in the aeration basin. There are two types of digestion most commonly practiced, mesophyllic and thermophyllic digestion. The mesophyllic digester operates at 85°F to 100° F (30°C to 38°C) to and thermophyllic digestion occurs at 120° F to 135° Fo (49°C to 57°C). Despite the higher temperatures, the average residence time for solids in the reactor is often high, well over 60 days. For comparable phytoremediation digestion the temperature considerations alone would indicate that the reaction rates are one fourth to one eighth of what happens in a digester.

A forthcoming paper by Alpesh Gohil and George Nakhla to be published in the Water Environment Federation Research Journal cites their work with bean and tomato processing wastewaters using an UASB (Upflow Anarobic Sludge Blanket) reactor. This reactor is shown below in Figure 4.

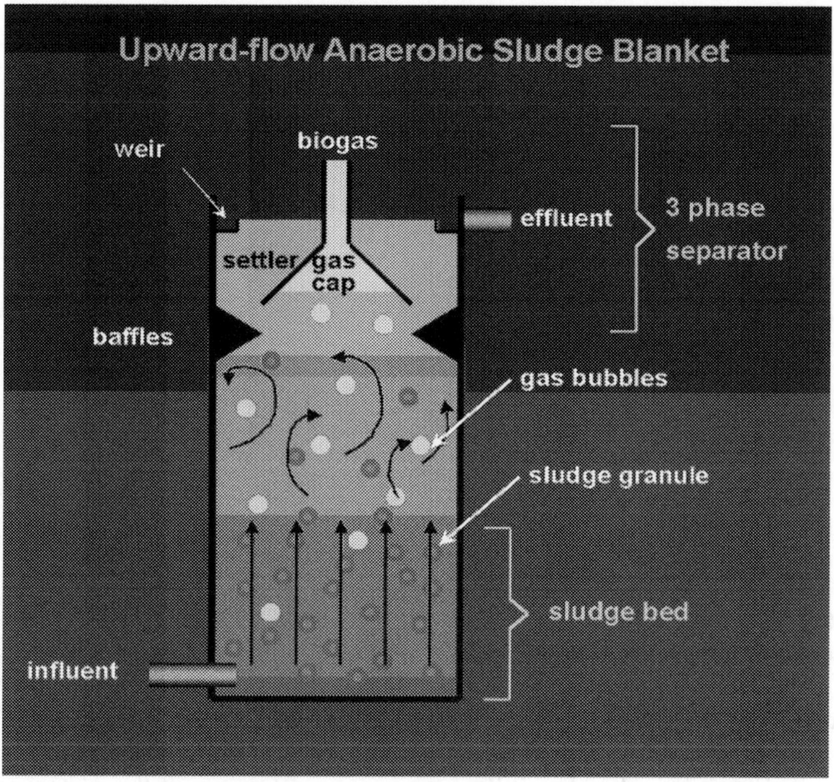

Figure 4. Drawing of UASB reactor from www.uasb.org.

The significance of this research showed that they could obtain excellent COD, BOD, TSS and NH3 removals in the effluent of their system with hydraulic residence times between 0.6 and 5 days and they obtained COD/N/P ratios of 167/3.3/1 and 127/2.8/1 which is substantially higher than the Metcalf and Eddy recommended 600: 5: 1 ratios for anaerobic reactors. Their work was performed at around 35oC + 2Co and their overall reactor treatment train looked like the following Figure 5.

The significance of this approach is that not only were the wastes treated at a very concentrated level using anaerobic and aerobic treatment, but the removals were extremely good. The implications for enhanced bioremediation suggest that some combination of aerobic and/anaerobic processes where nutrients are applied to the waste site and collected beneath the waste site could turn the entire waste site into an efficient bioreactor. These are interesting possibilities and the possibility of using a flooded system or other top down distribution system which recycles wastes from beneath the contaminated sites and returns it to the surface is an

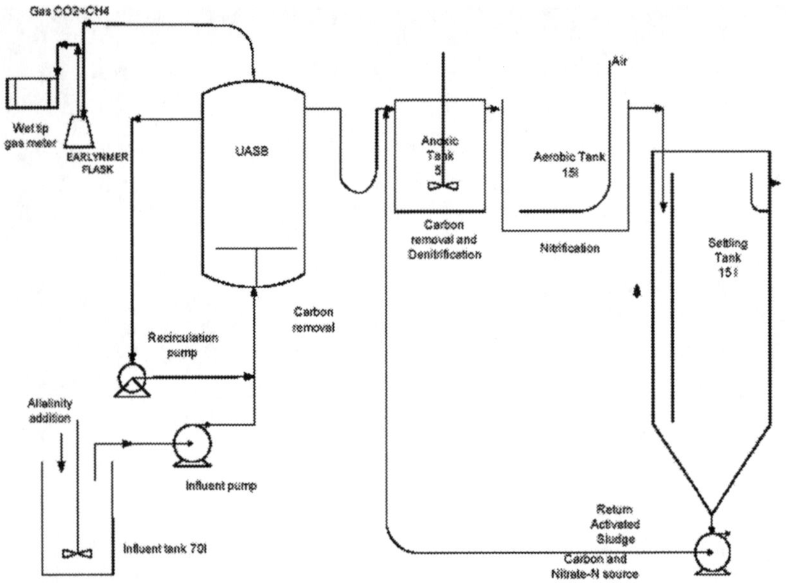

Figure 5. UASB Treatment Train for Tomato and Bean Wastes Note both aerobic and anaerobic treatment streams in sequence.

interesting possibility. Such a system could use horizontal wells for the collection of wastes from beneath the site, and when the major portion of the work was completed or it became necessary to convert the system to an aerobic system , the same horizontal wells which are used for collection of liquid could also be used for introduction of air into the formation to provide the aerobic environment one could find useful for further degradation.

5. Conclusion

The wastewater field has more than 100 years of treating difficult waste materials in an aqueous environment. There are sophisticated models for wastewater treatment which can be applied to the remediation field, and the methods and the approaches are also useful. The introduction of a C:N:P ratio or the introduction of various simple operating parameters has been cost effective, and may be applicable in the remediation field as well. From some of the new reactor designs and applications, it appears that some remediation problems may be those which have already been studied by the wastewater field, and that there are lessons to be learned as well as new and innovative approaches still remaining to be evaluated.

PHYTOREMEDIATION OF EXPLOSIVES

T. VANEK*[1], A. GERTH[2], Z. VAKRIKOVA[1],
R. PODLIPNA[1], P. SOUDEK[1]
1- Department of Plant Tissue Cultures, Institute of Organic Chemistry and Biochemistry, AS CR, Flemingovo nám. 2, 166 10, Praha 6, Czech Republic
2- BioPlanta GmbH, Deutscher Platz 5, D-04103 Leipzig, Germany

*To whom correspondence should be addressed.
Abstract- Results of experiments on Phytoremediation of Explosive and Energetic Compounds indicates that Phytoremediation is a promising technology for various levels of energetic compounds including TNT. This paper explores both the mechanisms and provides results of the work on phytoremediation.

Keywords: Remediation, Explosives, Phytoremediation

1. Introduction

The contamination of the environment by explosives, especially by nitroesters and nitroaromatics (NACs), is a worldwide environmental problem since enormous amounts of these compounds were produced during World War I and II. Most contaminated sites are located at ammunition factories and other places where these compounds were handled. This involved open detonation and burning of explosives at army depots, evaluation facilities, artillery ranges, and ordnance disposal sites (Rodgers and Bunce, 2001).

The most serious contamination was recognized as a result of production as well as use of 2,4,6-trinitrotoluene (TNT), mainly because of vast amount of water necessary for manufacturing of this compound made it necessary to locate the factories near natural sources of water – mostly rivers. The environment around large rivers is usually characterized by sand-rich soil that allows easy seepage of rainwater. NACs are therefore

N. Marmiroli et al. (eds.), Advanced Science and Technology for Biological Decontamination of Sites Affected by Chemical and Radiological Nuclear Agents, 209–225.

continually washed out from the contaminated soil and subsequently contaminate groundwater. As the sandy soil is not rich in humic substances, the capacity for binding of TNT and its derivatives is limited (Ahmad and Hughes, 2002; Wang et al., 2002). The persistency of NACs in the environment is due to resistance to biodegradation caused by their high toxicity for most living organisms (Lachance et al., 1999), low solubility in water (Ro et al., 1996), and consequently the absence of natural attenuation.

Over the last 100 years a variety of technological and medical applications have been found for nitrate esters, notably as high explosives and vasodilators. The most important nitrate esters are nitroglycerin (glycerol trinitrate, GTN) and pentaerythritol tetranitrate (PETN). Although useful therapeutically in low doses, nitrate esters and the resulting metabolites are generally toxic at higher levels, thus a significant environmental impact occurs. Wendt et al. (1978) found acute mammalian toxicity levels of 30 to 1300 mg/kg for GTN, whereas Urbanski (1984) determined LD50 of 1 mg/L for fish. These toxicity levels are of concern in wastewaters and soil contamination in the area of industrial plants manufacturing GTN and PETN, and waste containing GTN and nitrocellulose released handling or storage.

The long-established method for removing contamination has been excavation and subsequent incineration. This is applicable only in cases when the contamination is very high as the process is very costly. On the other hand, levels as low as 10 mg kg-1 would be considered to be high enough to require remediation in the U.S.A.

For such a low concentration less invasive alternative methods for removing contaminants from the environment are suggested (Snellinx et al, 2002; Gerth et al., 2003).

Generally, it is difficult to decide which technology should be applied at a certain concentration of TNT. In the case of phytoremediation, the upper limit of 1,000 mg kg^{-1} of soil has been proposed for efficient application (Gerth et al., 2003).

We concentrate our effort to phytoremediation of 2,4,6-trinitrotoluene (TNT) as the main contaminant of former and present ammunition factories and to nitroesters (PETN, NG) in the Czech Republic and in Germany. In our experiments, we studied the uptake, degradation and distribution of these compounds in plants and its degradation products in plant tissues and its subsequent utilisation for real application both for soil and waste-waters cleaning.

1.1. PHYTOREMEDIATION

Phytoremediation, one of the new biotechnologies, uses plants and their associated rhizospheric micro-organisms to remove, degrade, metabolise or detoxify contaminants including pesticides, metals, radionuclides, explosives, located in the soil, sediments, groundwater, surface water, and even the atmosphere (Chappell,1997). These plants can be herbs, shrubs or trees, and they may be able to accumulate organics and heavy metals high above the levels found in nature (Brown, 1995; Ma et al., 2000). Numerous mechanisms by which plants can remediate contaminated place exist – phytoaccumulation, phytoextraction, phytostabilization, phytotransformation, phytovolatilization and rhizodegradation.

Phytoremediation technology has two major advantages: it is relatively inexpensive and associated with minimal environmental disturbance. For these reasons phytoremediation has a high public acceptability.

Worldwide-established phytoremediation-based systems appear to adequately and efficiently remove pollutants from various matrices at a comparatively low cost. However, such successes have been achieved against a background of limited, formalized knowledge of the mechanisms involved, so that a more systematic approach concerning the selection of plants and optimization of remediation processes is urgently required (Schwitzguebel and Vaněk 2003). Drawbacks and limits of the phytoremediation technique may actually be seen in the comparatively long time required for the completion of the process, in the limitation to surface soils (root penetration depth), climatic factors modulating the viability of the plants and thus the remediation success, the toxicity of the pollutant and its degradation products to the plants and the possible need for soil amendment. Therefore the space opens for using techniques of molecular-biology and genetics to improve properties of (crop) plants and efficiently battle with the spread of contamination. Despite anti-GMO activities in EC genetically modified plants suited for phytoremediation represent new generation of GM plants. The genetically modified plants could be useful for efficient cleaning of wastewater as well.

1.1.1. *Phytoremediation of organic xenobiotics*

Phytoremediation of organic xenobiotics is generally based on mineralization or more frequently on degradation/transformation of the xenobiotics to environmentally less dangerous compounds, which are fixed in cell compartments or stored in the vacuole as soluble products or exuded back to the environment. Affection of the metabolic pathway by suppression or enhancement of expression of enzyme leading particular steps at degradation of xenobiotic by genetic modification might be a useful tool for efficient removing of xenobiotic from the environment.

1.1.2. *Phytoremediation of explosives*

1.1.2a. TNT. The TNT degradation pathway in plants is based on reduction of TNT nitrogroups (Larson et al., 1999; Rivera et al., 1998; Scheidemann et al., 1998; Vanderford et al., 1997). The reaction leads to the formation of aminodinitrotoluenes. Hydroxylamino- and amino derivatives of TNT were found to be transient intermediates which were transformed into different compounds, mainly conjugates or polymers. As an alternative degradation pathway, oxidation processes on a methyl group were observed (Bhadra et al., 1999b; Vanek et al., 2003). TNT degradation was studied using radiolabeled [^{14}C]-TNT (Bhadra et al., 1999a; Sens et al., 1999). Characterization and identification of TNT degradation products is difficult due to the insoluble/inextricable fractions bound in the plant cells. In the roots of wheat (*Triticum aestivum*) 27% of total activity was determined to be bound in lignin structures, while 57% of total activity was found in the cell wall (Sens et al., 1999).

Arabidopsis thaliana NADPH:thioredoxin reductase (TR, EC 1.6.4.5) has been found to catalyse reduction of nitroaromatics. The reduction is processed at a noncatalytic site of the enzyme. An important role may be played by glycosyltransferases, which were found to catalyse conjugation of 4-nitrophenol with mono- and disaccharides to form *O*-glycosides (Malcherek et al., 1998). Conditions of conjugation of hydroxylamino- or amino derivatives of polynitroaromatic compounds like TNT by *N*-glycosyltransferases still must be elucidated. Results on radioactivity distribution in bean roots suggest the formation of conjugates, mostly in lignin, cellulose and pectin fractions (Bhadra et al., 1999a). The formation of TNT conjugates with saccharides and polysaccharides might therefore be expected as confirmed by Vila et al. 2005.

The enzymatic system participating in degradation of TNT or other nitroaromatics in plants has not yet been sufficiently characterised. Based on recent results, TNT is not metabolised by highly specific nitroreductases, which would be purposefully synthesized by plants, but by constitutive enzymes with nitroreductase activity (Nepovim 2005b). This assumption is in agreement with results presented by Ekman (2005) showed an increase of the expression of a couple of enzymes in root mRNA due to the induction caused by explosives.

1.1.2b. Nitroesters. All reports of degradation of nitrate esters by prokaryotic and eukaryotic systems involve sequential denitration steps, resulting in multiple partially denitrated products shown in Figures 6-4 and 6-5.

All bacteria where nitrate ester degradation has been characterized have very similar enzymes. The enzymes catalyze the nicotinamide cofactor-dependent reductive cleavage of nitrate esters that produces alcohol and nitrite. Purification of the PETN reductase from *Enterobacter cloacae* yielded a monomeric protein of around 40 kilo Daltons, which required NADPH as a co-factor for activity. Similar enzymes were responsible for the nitrate ester-degrading activity in *Agrobacterium radiobacter* (Snape et al. 1997) – "nitrate ester reductase" – and in the strains of *Pseudomonas fluorescens* and *Pseudomonas putida* (Blehert et al. 1999) – "xenobiotic reductases". All utilize a non-covalently bound flavine mononucleotide as a redox cofactor.

By contrast, there are few reports concerning nitroester degradation in plants. Goel et al. (1997) demonstrate, that beetroot (*Beta vulgaris*) denitrate GTN, and the denitration was enhanced by expression of PETN reductase in transgenic seedlings, which also enhanced the denitration of glyceroldinitrate (GDN) to glycerolmononitrate (GMN). French et al. (1999) followed the same approach and developed transgenic tobacco plants (*Nicotiana tabacum*) that express PETN reductase to degrade nitrate ester explosives and TNT. Seeds from this transgenic plant were able to germinate and grow in media containing GTN and TNT levels toxic to the wild type tobacco. Recent data are summarized in Hannink et al. 2002

2. Materials and methods

2.1. SUSPENSION CULTURES

Rheum palmatum: Callus culture was induced from petiole of plant cultivated *in vivo* on the Murashige-Skoog (MS) medium (Murashige and Skoog, 1962) gelled by agar (8g/1), supplemented with naphthaleneacetic acid (NAA) (10 mg/l) and casein hydrolyzate (2 g/l). Callus subcultured every 4 weeks was transferred into the liquid medium and filtered through a sieve (mesh 3 mm).

2.2. *IN VITRO* WATER PLANTS

The plant species *Phragmites australis* and *Typha latifolia* were obtained from surface-sterilized seeds whereas *Carex gracillis* and *Juncus glaucus* were initiated from apical meristem. The seeds were treated by a combination of sodium hypochlorite (1%, 10 min, *in vacuo*) and mercuric chloride (10 ppm, 5 min). The washed seeds were sown on the hormone-free MS medium (Murashige et al. Skoog, 1962). The meristem was excised from the surface-sterilized apical part of plant. The plants were aseptically cultivated in 150 ml non-mixed containers without any solid support with a light period of 16 hours on basal MS medium supplemented by vitamins: Glycine (2 mg/l), Myo-inositol (100 mg/l), Nicotinic acid (0.5 mg/l), Pyridoxine HCl (0.5 mg/l), Thiamine HCl (1 mg/l), by phytohormone BAP (5 mg/l) and by Saccharose (30 g/l) at 25°C. During the experiments 20 ml of liquid medium was used and the inoculum was about 7 g of plants. The base of plants (1 cm) was permanently immersed and the medium was not changed during the treatment. Roots were induced on the hormone – free medium during a 4-5 weeks cultivation period.

Sterility of the whole system before, during and after experiments was checked by plating out the medium onto agar plates.

2.3. CHEMICALS

2,4,6-trinitrotoluene, glyceroltrinitrate and pentaerytritoltetranitrate were gained from ammunition factory Synthesia (Pardubice, Czech Republic). Degradation products (2ADNT and 4ADNT; 2,4DANT and 2,6DANT; TNB, GDN, GMN) were supplied from Department of Explosives (University of Pardubice).

2.4. HPLC ANALYSIS

Analytical instrumentation consisted of binary gradient pumps Deltachrom SDS 020 & 030 (USA), a mixer (SunChrom GmbH, BRD), an injection valve Rheodyne 7725 (Rheodyne, USA) and a PDA detector MD-1510 (Jasco, Japan). Data were processed by Borwin PDA program and the concentration of explosives were calculated from peak area at wavelength 230 nm. Analyses were performed on a stainless steel column (250 x 4 mm ID) packed with SiC_{18} reverse phase Biospher 7μm size (Labio Ltd., Czech Rep.). A linear gradient of mobile phases (10% - 100% MeOH within 40 min) was applied.

An uptake of TNT was determined in a medium and in cells. Medium was filtered through a microfilter (0.2 μm) and 20 μl was injected. The content of explosives in cells was determined from acetone extract.

2.5. LC-MS

HPLC - Beckman 125 binary gradient pumps, 168 “diode-array” detector, 507 autosampler MS – “Ion-trap” mass spectrometer Finnigan LCQ equipped by APCI (atmospheric pressure chemical ionization), data analyzed in negative mode, spectra confirming found compounds were obtained from tandem mass spectromectry (MS/MS).

2.6. IDENTIFICATION

Identification of explosives and respective degradation products was done by comparing UV and MS spectra of known compounds from HPLC and LC-MS, respectively as well as by comparing their retention times with standards.

2.7. PHYTOTOXICITY EXPERIMENTS

Five different concentrations (0, 25, 50, 75, 100 mg/l) of explosives were added to the cultivation medium from stock solution of TNT (100 mg/ml, DMSO). The experiment were done in triplicates for particular concentration and harvest-day. The concentration which inhibits growth of suspension culture by 50% (IC50), was calculated from GVs, which corresponded to the end of exponential phase of growth and which differed for particular species. The GVs for different concentration of explosives were plotted and IC50 calculated.

2.8. DEGRADATION AND UPTAKE EXPERIMENTS

Explosives were added into the cultivation medium at a concentration of 50 mg/l at the first part of experiment and repeatedly at two different concentrations (10 and 50 mg/l) at the other part of experiment.

3. Results

3.1. DEGRADATION OF EXPLOSIVES BY SUSPENSION CULTURES

3.1.1. *TNT*

The uptake of TNT by *R. palmatum* was determined over 24 hours. The initial concentration of TNT decreased to 50% within half an hour. No TNT could be detected after 6 hours (Fig. 1). The exponential shape of the curve suggests that enzymatic degradation of TNT is undeniable (Nepovim et al. 2004).

A production of aminodinitrotoluenes and diaminonitrotoluenes was observed in culture of *R. palmatum*. The changes of their concentration in media during cultivation is displayed on the Fig. 1. In the cells, the same products as in media were identified at the end of experiment in the same ratio.

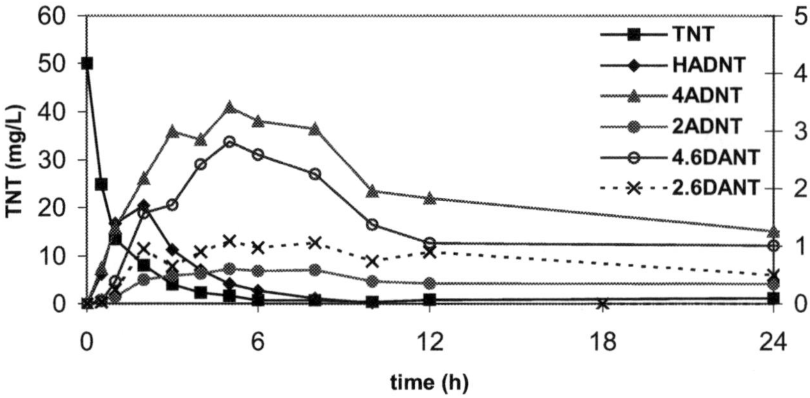

Figure 1. Time-course of trinitrotoluene degradation by *Rheum palmatum* cells. TNT = trinitrotoluene, HADNT = hydroxyaminodinitrotoluene, 4ADNT = 4-aminodinitrotoluene, 2ADNT = 2-aminodinitrotoluene, 4,6DANT = 4,6-diaminonitrotoluene, 2,6DANT = 2,6-diaminonitrotoluene.

This observation support the degradation pathway suggested by Burken (2000) and Rivera (1998) (Fig. 2).

Alternative degradation *via* TNB in plants was described by Vanek et al. 2003 using *in vitro* cultivated poplar.

3.1.2. *GTN and PETN*

Degradation of nitroesters by *Rheum palmatum* tissue culture proceeded along the pathway shown in Figures 3 and 4. The concentration of GTN was decreased to 39 % of the initial concentration of 50 mg/L within ten days, while the same concentration of PETN was totally transformed. Vanek et al. 2003). From the analysis of products and stoichiometry of both processes, glycerol and pentaerythritol were formed as end products, which can eventually be used by plant cells as a carbon source (Maestri et al. 1991).

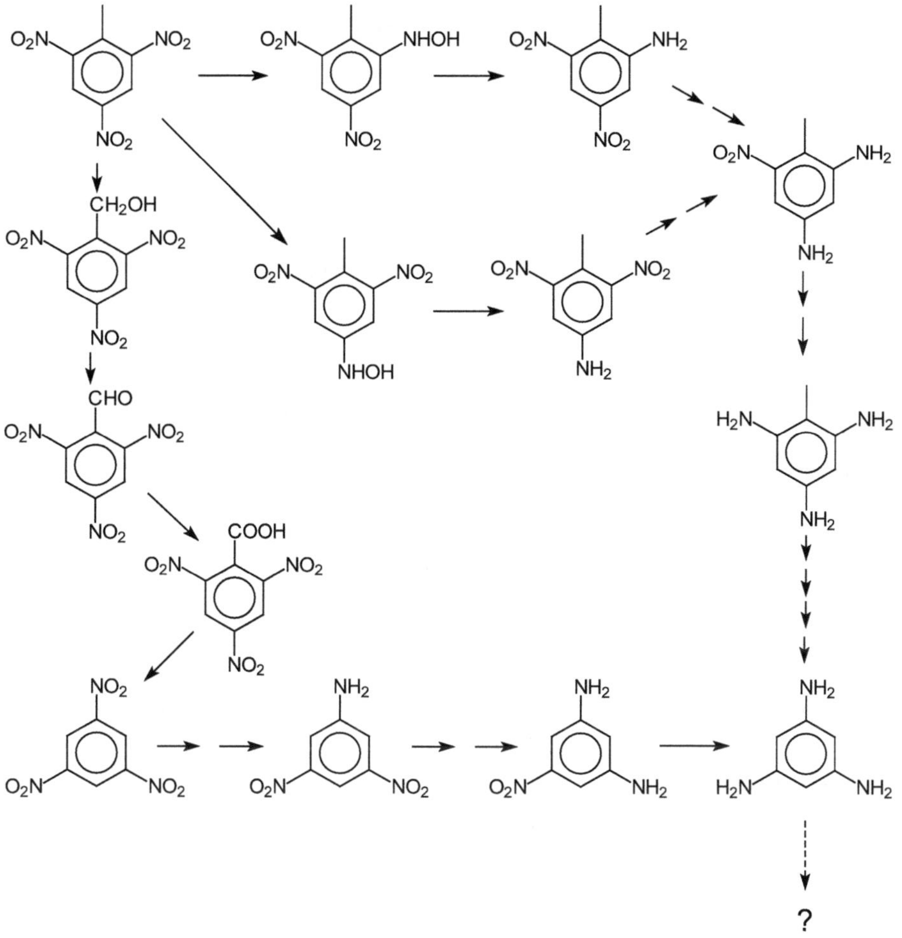

Figure 2. Proposed reduction and oxidation pathways of TNT in plants (Rivera et al. 1998).

The achieved results give us valuable information about kinetics of explosives degradation *in R. palmatum* plant cells and about soluble metabolites formed during degradation. This results can be extended for whole plants too – it is generally known, that the spectrum of resulting metabolites in plants and plant cell cultures is in principle identical, though quantitative differences may occur (Harms 1992).

3.1.3. *Degradation of TNT by water plants*

The disappearance of TNT from the medium is connected with the uptake of TNT into the plant tissues and its transformation into degradation products. About 80% of TNT was taken up from the medium in 3 days and

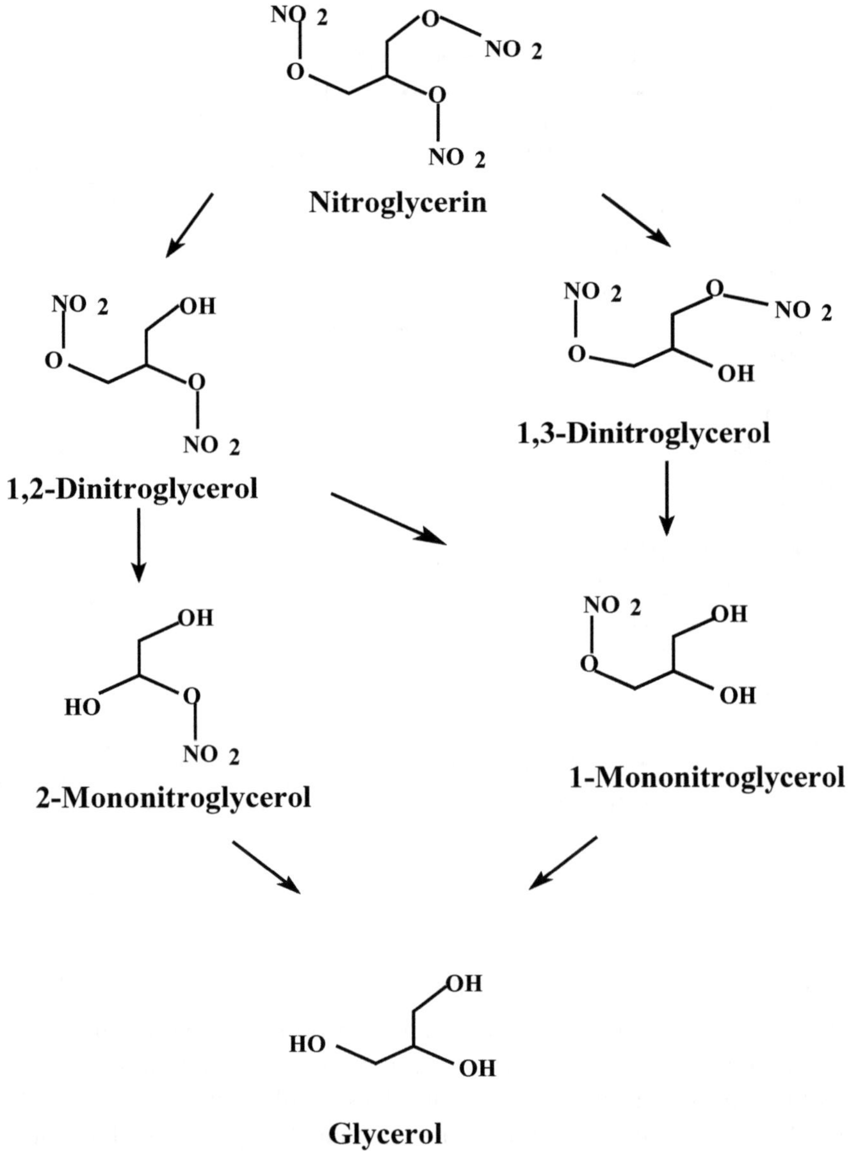

Figure 3. Degradation pathways of nitroglycerin.

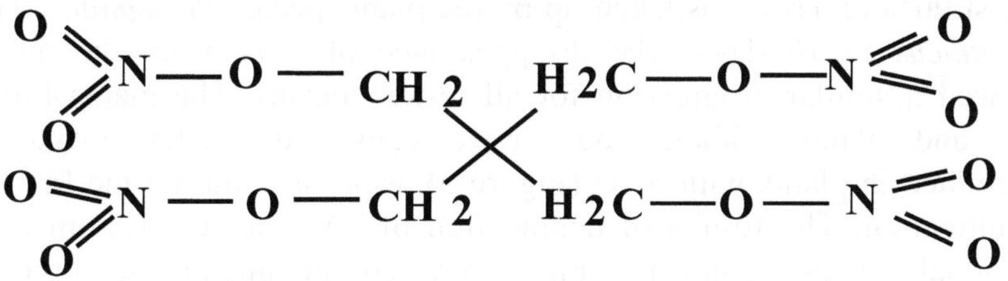

Pentaerythritol tetranitrate

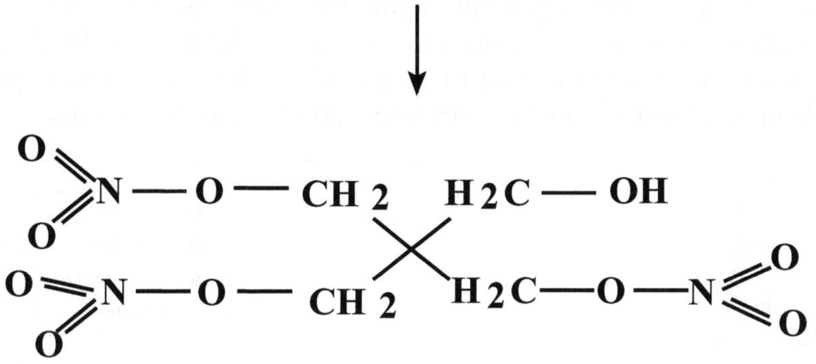

Pentaerythritol trinitrate

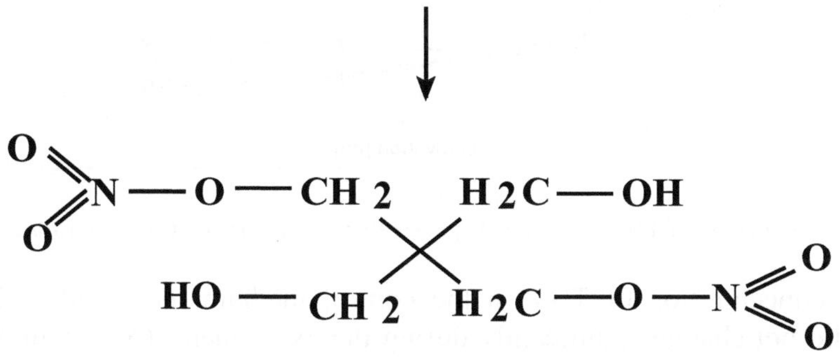

Pentaerythritol dinitrate

Figure 4. Scheme of pentaerythritol tetranitrate degradation - The degradation continued via pentaerythritol mononitrate to pentaerythritol, which might be utilized as a carbon source.

at least 90% of TNT was taken up by the plant species *Phragmites*, *Typha* and *Juncus* in 10 days. The disappearance of TNT from the medium followed a similar progression for all tested species. The plants forming roots and plants without roots were compared. Faster uptake was determined in plants with roots (Figure 5), especially during the first days of cultivation. The results of degradation of TNT in the medium by the individual species are shown in Figure 6, where the time profile of the TNT degradation products 4-ADNT and 2-ADNT is presented. The formation of monoaminoderivatives ranged from 1-10% of the initial concentration after one day of cultivation depending upon the plant species. The highest concentration of 4-ADNT remaining in the medium after 10 days of cultivation was 5.4% in the case of *C. gracillis,* whereas the other species left only up to 2% of aminodinitrotoluenes (ADNTs) in the medium.

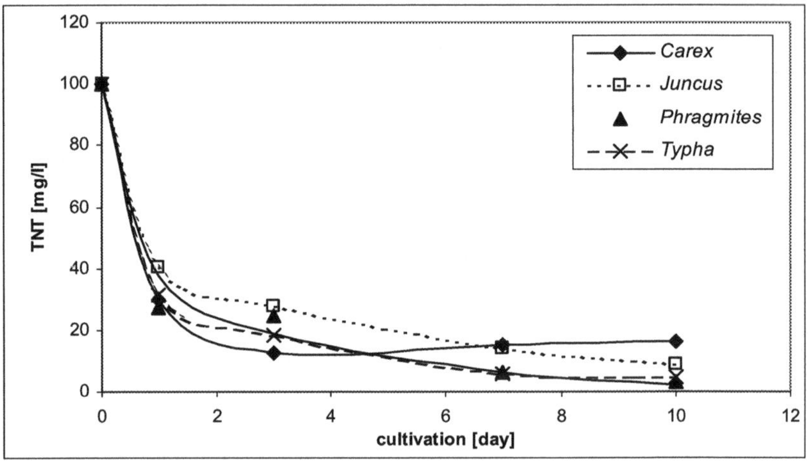

Figure 5. Degradation of TNT by the plant species *Phragmites*, *Typha Carex* and *Juncus*.

The concentration of TNT in the control medium incubated without plants did not change significantly during the experiment (Nepovim et al. 2005).

3.2. REAL APPLICATION

Emergent plants (helophytes) showed a potential for removal of TNT from contaminated water under *in vitro* conditions with small differences in the formation of the major degradation products – monoaminodinitrotoluenes. Most of TNT degradation products (using ^{14}C-radiolabelled TNT) were localized in the roots of reed (53% of total radioactivity) as insoluble compounds (33% of total radioactivity) (Nepovim et al. 2005).

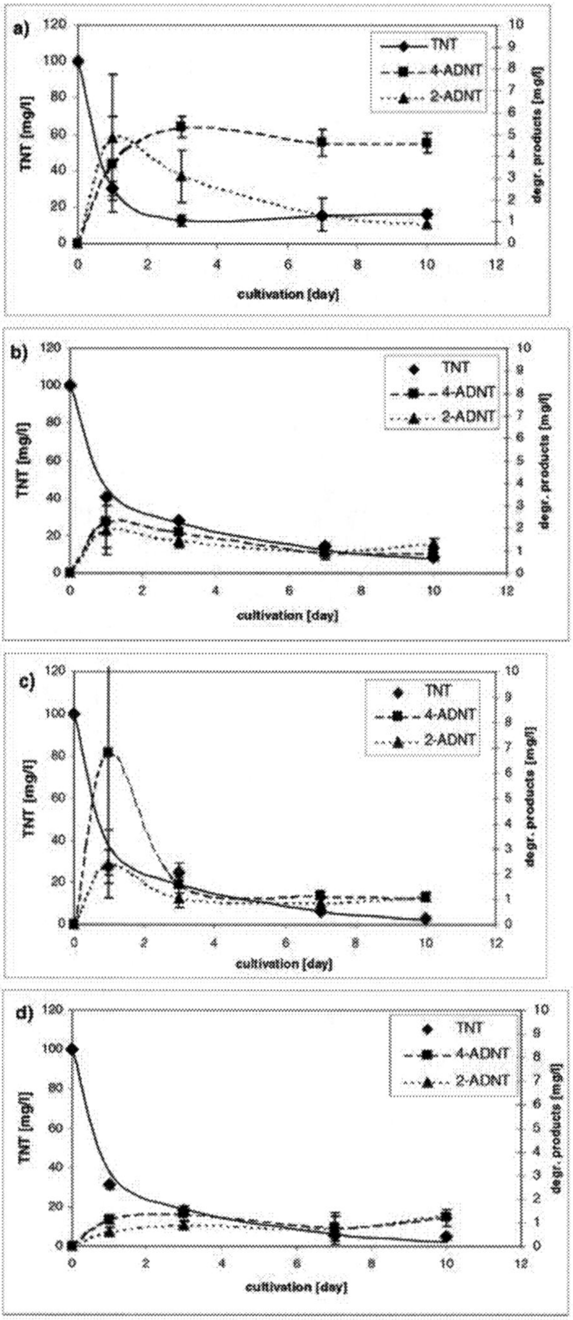

Figure 6. The concentration of TNT and aminodinitrotoluenes (4-ADNT, 2-ADNT) in medium treated by *C. gracillis* (a), *J. glaucus* (b), *Ph. australis* (c), *T. latipholia* (d) for 10 days.

Based on the above-mentioned results, *Ph. australis* seems to be the best candidate for practical application. This idea is supported by the fact that it forms prolific biomass necessary for efficient biodegradation and is able to grow almost everywhere.

The possibility of this approach was tested during pilot-scale installation of constructed wetland for cleaning of waste-waters containing nitroester explosives in producing factory, in model arrangement consisted of plastic containers $1 m^3$ volume each (Fig. 7) with *Phragmites*. In this arrangement plants were able to clean water containing 270 mg/l of nitroesters and their by-products during 30 days (Fig. 8). Achieved results were verified and confirmed real-scale application (Vanek et al 2005) using *Phragmites* and *Typha* plants.

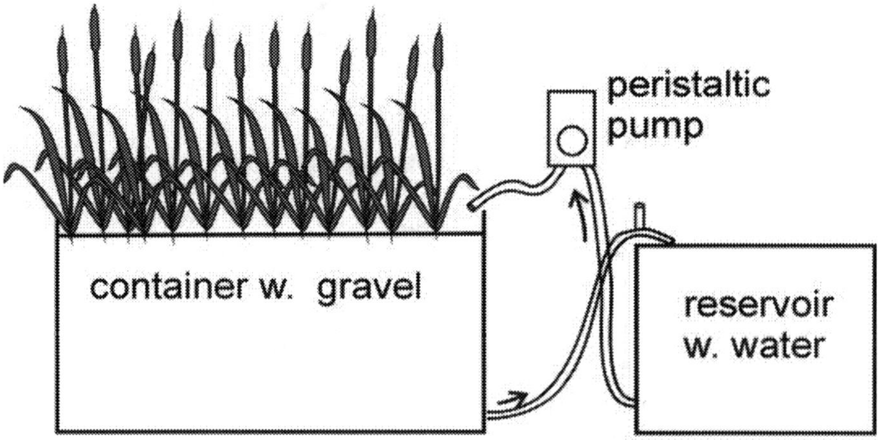

Figure 7. Scheme of model constructed wetland.

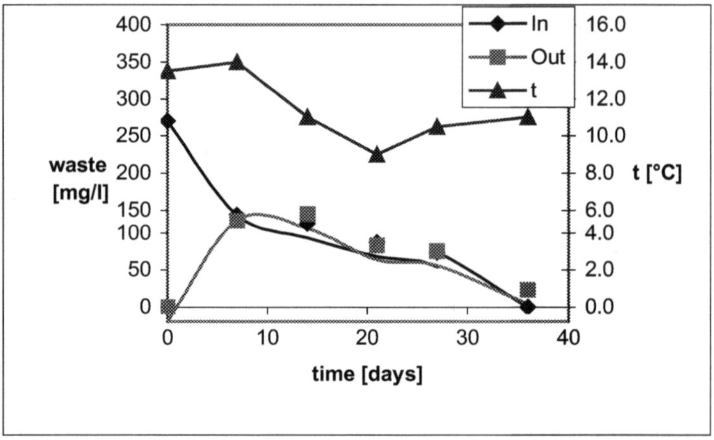

Figure 8. Degradation of nitroesters and their by-products in industrial waste-waters.

Acknowledgements

This work was supported by COST 859.10

References

Ahmad, F., and Hughes J. B., 2002, Reactivity of partially reduced arylhydroxylamine and nitrosoarene metabolites of 2,4,6-trinitrotoluene (TNT) toward biomass and humic acids, *Environ. Sci. Technol.* **36**:4370-4381.

Bhadra, R., Spanggord, R. J., Wayment, D. G., Hughes, J. B., and Shanks, J. V., 1999a, Confirmation of conjugation processes during TNT metabolism by axenic plant roots, *Environ. Sci. Technol.* **33**:3354-3361.

Bhadra, R., Wayment, D. G., Hughes, J. B., and Shanks, J. V., 1999b, Characterization of oxidation products of TNT metabolism in aquatic phytoremediation systes of *Myriophyllum aquaticum*, *Environ. Sci. Technol.* **33**:446-452.

Blehert, D. S., Fox, B. G., and Chambliss, G. H., 1999, Cloning and sequence analysis of two *Pseudomonas* flavoprotein xenobiotic reductases, *J. Bacteriol.* **181**:6254-6263.

Brown, K.S., 1995,. The green clean, *Bioscience* **45**:579–582

Burken, J. G., Shanks, J. V. and Thompson, P. L., 2000, Phytoremediation and plant metabolism of explosives and nitroaromatic compounds, in: *Biodegradation of Nitroaromatic Compounds and Explosives*, J. C. Spain, J. B. Hughes, H. J. Knackmuss, eds., Lewis Publishers, Boca Raton, pp. 240-275.

Chappell, J., 1997, Phytoremediation of TCE Using *Populus*, Status report prepared for the US EPA Technology Innovation Office under a National Network of Environmental Management Studies Fellowship, pp.1-38.

Courbot, M., Craciun, A., Heudiard, A., Kieffer, P., Roosens, N., Saumitou-Laprade, P., Willems, G., and Verbruggen, N., 2004, Identification of genes involved in Cd tolerance in Cd hyperaccumulation, COST Action 859, 1st WG2 Workshop – Exploiting "-omics" approaches in phytotechnologies, p. 11.

Ekman, D. R., Wolfe, N.L., and Dean, J.F.D., 2005, Gene expression changes in *Arabidopsis thaliana* seedling roots exposed to the munition hexahydro-1,3,5-trinitro-1,3,5-triazine. *Environ. Sci. Technol.* **39**:6313-6320.

French, C. E., Rosser, S. J., Davies, G. J., Nicklin, S. and Bruce, N. C., 1999, Biodegradation of explosives by transgenic plants expressing pentaerythritol tetranitrate reductase, *Nature Biotechnol.* **17**:491-494.

Gerth, A., Hebner, A., and Thomas, H., 2003, Natural remediation of TNT-contaminated water and soil, *Acta Biotechnol.* **23**:143–150.

Goel, A., Kumar, G., Payne, G. F., and Dube, S. K., 1997, Plant cell biodegradation of a xenobiotic nitrate ester, nitroglycerin, *Nature Biotechnol.* **15**:174-177.

Hannink, N. K., Rosser, S. J., and Bruce, N. C., 2002, Phytoremediation of explosives, *Crit. Rev. Plant Sci.* **21**:511-538.

Harms, H. H., 1992, In-vitro systems for studying phytotoxicity and metabolic fate of pesticides and xenobiotics in plants. *Pestic. Sci.* **35**:277-281.

Lachance, B., Robideux, P. Y., Hawari, J., Ampleman, G., Thiboutot, S., and Sunahara, G. I., 1999, Cytotoxic and genotoxic effects of energetic compounds on bacterial and mammalian cells *in vitro*, *Mutat. Res.* **444**:25-39.

Larson, S. L., Jones, R. P., Escalon, L., and Parker, D., 1999, Classification of explosives transformation products in plant tissue, *Environ. Toxicol. Chem.* **18**:1270-1276.

Ma, L. Q., Tu, C., Kennelley, E. D., and Komar, K. M., 2000, Phytoremediation of arsenic contaminated soils and wastes, Annual meetings abstracts, American Society of Agronomy, Minneapolis, Minn. November, pp. 5–9.

Maestri, E., Restivo, F. M., Gulli, M., and Tassi, F., 1991, Glutamate-dehydrogenase regulation in callus-cultures of *Nicotiana plumbaginifolia* - effect of glucose feeding and carbon source starvation on the isoenzymatic pattern, *Plant Cell Environ.*, **14**:613-618.

Malcherek, K., Breuer, J., Schumphan, I., and Schmidt, B., 1998, Metabolism of 4-nitrophenol in asepticaly cultivated plants of the species wheat (*Triticum aestivum* L.), soybean (*Glycine max* L.), wild oat (*Avena fatua* L.) and corn cockle (*Agrostemma githago* L.). *J. Plant Physiol.* **153**:192-199.

Murashige, T., and Skoog, F., 1962, A revised medium for rapid growth and bio assays with tobacco tissue cultures, *Physiol. Plant.* **15**:473-497.

Nepovim, A., Hebner, A., Soudek, P., Gerth, A., Thomas, H., Smrcek, S., and Vanek, T., 2005, Phytoremediation of TNT by selected helophytes, *Chemosphere* **60**:1454–1461.

Nepovím, A., Hubálek, M., Podlipná, R., Zeman, S., and Vaněk T., 2004, *In vitro* degradation of 2,4,6-trinitrotoluene by plant tissue culture of *Solanum aviculare* and *Rheum palmatum,. Eng. Life Sci.* **24**:46-49.

Nepovím, A., and Vaněk T., 2005b, Soapwort oxophytodienoate reductase participates in the degradation of trinitrotoluene, in prep.

Rivera, R., Medina, V. F., Larson, S. L., and McCutcheon, S. C., 1998, Phytotreatment of TNT-contaminated groundwater, *J. Soil Contamin.* **7**:511-529.

Ro, K. S., Venugopal, A., Adrian, D. D., Constant, D., Qaisi, K., Valsaraj, K. T., Thibodeaux, L. J., and Roy D., 1996, Solubility of 2,4,6-trinitrotoluene (TNT) in water, *J. Chem. Eng. Data* **41**:758-761.

Rodgers, J. D., and Bunce, N. J., 2001, Treatment methods for the remediation of nitroaromatic explosives, *Wat. Res.* **35**:2101-2111.

Scheidemann, P., Klunk, A., Sens, C., and Werner, D., 1998, Species dependent uptake and tolerance of nitroaromatic compounds by higher plants, *J. Plant Physiol.*, **152**:242-247.

Schwitzguebel, J. P., and Vaněk, T., 2003, Some fundamental advances for xenobiotic chemicals, in: *Phytoremediation: Transformation and control of contaminants,* S. C. McCutcheon, J. L. Schnoor, eds., John Wiley & Sons, Inc., Hoboken, pp. 123-157.

Sens, C., Scheidemann, P., Klunk, A., and Werner D., 1998, Distribution of ^{14}C -TNT and derivatives in different biochemical compartments of *Phaseolus vulgaris*, *Environ. Sci. Pollut. Res.* 4:202-208.

Sens, C., Scheidemann, P., and Werner, D., 1999, The distribution of ^{14}C-TNT in different biochemical compartments of the monocotyledonous *Triticum aestivum. Environ. Pollut.* **104**:113-119.

Snape, J. R., Walkley, N. A., Morby, A. P., Nicklin, S., and White G. F., 1997, Purification properties and sequence of glycerol trinitrate reductase from *Agrobacterium radiobacter, J. Bacteriol.* **179**:7796-7802.

Snellinx, Z., Nepovim, A., Taghavi, S., Vangronsveld, J., Vanek, T., and van der Lelie, D., 2002, Biological remediation of explosives and related nitroaromatic compounds, *Environ. Sci. Poll. Res. Int.* **9**:48-61.

Urbanski, T., 1984, *Chemistry and Technology of Explosives*, Pergamon Press Ltd., Oxford.

Vanderford, M, Shanks, J. V., and Hughes, J. B., 1997, Phytotransformation of trinitrotoluene (TNT) and distribution of metalic products in *Myriophyllum aquaticum*, *Biotechnol. Lett.* **19**:277-280.

Vanek, T., Nepovim, A., Podlipna, R., and Vavrikova Z., 2005, Czech. Pat. Appl.

Vanek, T., Nepovim, A., Podlipna, R., Zeman, S., and Vagner, M., 2003, Phytoremediation of selected explosives, *Water Air Soil Pollut. Focus* **3**:259-267.

Vila, M., Pascal-Lorber, S., Rathahao, E., Debrauwer, L., Canlet, C., and Laurent, F. 2005, Metabolism of [C-14]-2,4,6-trinitrotoluene in tobacco cell suspension cultures, *Environ. Sci. Technol.* **39**:663-672.

Wang, C. J., Thiele, S., and Bollag, J. M., 2002, Interaction of 2,4,6-trinitrotoluene (TNT) and 4-amino-2,6-dinitrotoluene with humic monomers in the presence of oxidative enzymes. *Arch. Environ. Contam. Toxicol.* **42**:1–8.

Wendt, T. M., Cornell, J. H., and Kaplan, A. M., 1978, Microbial degradation of glycerol nitrates, *Appl. Environ. Microbiol.* **36**:693-699.

PART 4

REGULATORY ISSUES

GERMAN LAW FOR THE PREVENTION, IDENTIFICATION, INVESTIGATION, RISK ASSESSMENT AND REMEDIATION OF CONTAMINATED SITES AND HARMFUL SOIL CHANGES

WOLF-UWE MARR*
Bundesministerium der Verteidigung, WV IV 2, Postfach 1328, D53003 Bonn, Germany
Email: wolfuwemarr@bmvg.bund400.de

*To whom correspondence should be addressed.

Abstract- Protection of the environment is as much a public responsibility as the duty of every citizen. State authorities issue appropriate regulations and monitor their performance. The German Basic Law (Grundgesetz) declares protection of the environment to be a federal objective and the responsibility of the States. The Federal Soil Protection Act (Bundes-Bodenschutzgesetz, BBodSchG) and the Federal Water Act (Wasserhaushaltsgesetz, WHG) as well as the Federal Soil Protection and Site Remediation Ordinance (Bundes-Bodenschutz- und Altlastenverordnung, BBodSchV) give a framework for the States' regulatory issues. The main objective of the BBodSchG is to protect in a sustainable way the functions of the soil or to restore them. Damage to and contamination of the soil should be avoided, and where it is detected, it must be remedied. The landowner and the main user must prevent damage or contamination. The landowner, the main user, and the person who caused the damage/contamination are responsible for remediation. Where there is reasonable cause to suspect the occurrence of soil damage /contamination, the responsible local authority may oblige one of the above-mentioned persons to monitor contaminant content, perform a risk assessment and possibly apply remediation. Two groups of contaminated sites have been defined: (1) where contamination was caused by a former use of the site different from its current use (Altlasten), and (2) where contamination is caused by the actual use of the site (schädliche Bodenveränderung). But in either case, corresponding measures must be initiated. The Federal Water Protection Act (Wasserhaushaltsgesetz) regulates the protection of ground and surface water and its use. Landownership does not include the right to

N. Marmiroli et al. (eds.), Advanced Science and Technology for Biological Decontamination of Sites Affected by Chemical and Radiological Nuclear Agents, 229–256.

use ground and surface water or the run-off of grey water. All must take care to maintain good water quality and usability. Public good (Wohl der Allgemeinheit) and individual usability are the predominant objectives. The German regulatory system is discussed in detail; and the author's personal view of the meaning for biological remediation techniques, which may differ from the official view of the government, is introduced.

1. Introduction

Most regulations arise because of need, so those covering the use, treatment, distribution and conservation of water and water resources are very old and have been amended several times to keep abreast of current requirements. As water has traditionally been a matter of common use, it is often considered as common property, and private ownership is very restricted. Under German law, ground- and surface-water do not belong to the land, and if the landowner wishes to use it, an application to do so must be transferred to the water office, for which, in some situations, a payment is required. Water regulations are based on the German Water Management Act and the European Union (EU) Water Framework Directive. The regulations covering soil are very different. Ownership of land implies personal property, so that soil can be managed according to the owner's needs and wishes. Limitations only apply when the land presents a danger to others. This basic concept has been incorporated into law, making the sustainable management of land (including the soil) a direct legal duty.

Industrialisation brought many changes to the industrial and in the agricultural sectors, and even residential areas have become affected by many kinds of pollution. Modern agricultural machinery causes soil compaction, and erosion has increased to dangerous levels. In addition, many hazardous substances have been introduced into the soil, such as fertilizers, insecticides, herbicides and sewage sludge. With the rise of industry and mass production, the area of contaminated land increased, exacerbated by war, which brought chemical and other weapons, exploded ordnance, and explosives production sites. Many of the latter were bombed in Germany whilst still in production, leading to the escape of hydrocarbons and cleaning liquids. Since the war, residential developments, retail centres and industrial plants have expanded into rural areas, covering land formerly occupied by forest and agriculture. As a result, it has become necessary to enforce soil protection and soil conservation, first by amending existing laws of land use planning, emission protection, waste disposal and others, and finally by developing a soil protection law. In 1984 the German Federal Government presented a concept of soil protection, and in 1989 the Federal

Council of Experts introduced proposals for the detection, investigation and remediation of contaminated sites. These laid the foundation for establishing a Federal Soil Protection Act. In the meantime, in the absence of a federal Act, the States established their own varied regulations, which has complicated the task of consultants, landowners and remediating companies.

At that time, the decision as to whether measures need to be carried out was based mainly on proposals made in various scientific publications or derived from them (such as the Netherlands list, the Hamburg list, Kloke values and others), and differed depending on the local authority. The best advice was to consult the environmental office before starting any work. The decisions of the environmental authority were based mainly on police law, which required measures to be undertaken where there was immediate danger to the public (i.e., man, water, soil or air). The resulting confusion of different regulations needed to be unified by federal law in order to define both legal limits of hazardous substance concentrations and legal methods to be used for investigation, assessment and remediation of contaminated sites. The German Federal Soil Protection Act (BBodSchG) was enacted in 1998, and more detailed regulations and technical standards followed in 1999 in the Federal Soil Protection and Contaminated Sites Ordinance (BBodSchV). The States are permitted to establish more detailed regulations within the framework of the federal law. This would have marked the end of the legal story had not the EU declared the need for a European Thematic Soil Strategy and the announcement of its intention to develop an EU Soil Framework Directive (SFD). Thus amendments to the regulations will have to continue.

2. General aspects of regulations covering investigation, assessment and remediation

The EU directives have set the general goals for soil and water protection, and suggest the regulations necessary to attain these objectives.

2.1. THE WATER FRAMEWORK DIRECTIVE

The main objective of this directive is to attain a sustainable supply of natural water, to improve its quality over current standards, and to guarantee sufficient supply. This is to be achieved by detailed monitoring in Member States and in a steady improvement by prescribed measures. The regulation load will therefore increase. Based on this directive, the national law in Germany has been amended.

2.2. EU THEMATIC STRATEGY ON SOIL PROTECTION AND THE PROJECTED SOIL FRAMEWORK DIRECTIVE

Soil threats expected to be addressed by the directive are:

- contamination
- sealing
- erosion
- organic matter decline
- salinisation
- compaction
- floods and landslides
- biodiversity loss

The soil framework directive should be presented by the end of 2005. At the current time it remains uncertain which of the above soil threats will be regulated by the EU and which will be the responsibility of member States. However, the number of regulations is bound to rise. Furthermore, new measures will probably have to be paid for by private owners based on their duty of care. Only orphan sites may attract public funding.

3. German regulations

German regulations covering environmental protection, specifically the identification, investigation, assessment and remediation of polluted soils and waters, are based on the German Basic Law (Grundgesetz, GG), the Federal Soil Protection Act and the Federal Water Act (Wasserhaushaltsgesetz), and derived Ordinances. Article 20 of the GG guarantees the protection of natural foundations of life to future generations, which means a lifelong environment guarantee enforced by law, government and the courts. The GG distributes responsibilities between the Federation and the States (Länder). Thus responsibility for establishing the regulatory framework lies with the Federal Government, while the States may implement and supplement Federal regulations and act as the regulatory authority.

3.1. THE FEDERAL WATER ACT (WASSERHAUSHALTSGESETZ)

This Act was amended in 2001 respecting the requirements of the EU water framework directive. It regulates the use and protection of surface water, coastal water and groundwater, but not waterways. Water is considered as a part of nature and needs to be preserved as an environment for plants and animals. It must be used sustainably. Private ownership does not authorize uses restricted in this Act or any alteration by construction. The law defines "use of water" (Article 3 WHG) as the drawing and deflecting of surface water, change of water surface or water level, excavating solids from the water beds, discharge of substances into water, and measures which are likely to cause significant degradation of chemical and biological characteristics. For any such use, permission from the competent authority is needed, and this will define any further conditions of use. The decision of the water authority is governed by considerations of the public good. The specific Federal Ordinance "Requirements for the Discharge of Waste Water into Waters (Waste Water Ordinance – AbwV)" establishes special restrictions applying to the construction, land-use and traffic of water conservation areas (Article 19 WHG).

The water offices are involved in the planning and operation of remediation in many ways, both in issues surrounding the collection of surface water in a basin, and run-off (discharge) from a basin into the ground-, underground- or surface-water. The water office does not need to prove the actual presence of danger or damage to the water in order to justify the ordering of measures; it is sufficient that there is considered to be a risk that water quality may be damaged. Examples will be presented below. The text of the Act and Ordinance is available in German and English on www.bmu.de. The responsibilities within the water Act and soil protection Act are well defined; the soil office is the competent authority for soil protection and site remediation, even where water is involved, and it will give the orders and permissions, even if it has to consult the water office regarding specific requirements for water protection in cases where waters are concerned.

3.2. FEDERAL SOIL PROTECTION LAW

The Federal Soil Protection Act and Federal Soil Protection and Federal Contaminated Site Ordinance (BBodSchV) describe the main and primary regulations for the identification, investigation and assessment of contamination, and for the planning and operation of remediation. Every expert should be familiar with both the main rules and the details. Published comments on the Act and its Ordinance may be a help for better understanding.

3.2.1. *Federal Soil Protection Act (BBodSchG)*

The objective of the Act, as defined in Article 1, is "to protect or to restore the functions of the soil on a permanently sustainable basis. These actions shall include the prevention of harmful changes to the soil, the rehabilitation of the soil in contaminated sites and of waters contaminated by such sites; and precautions against negative soil impacts. Where impacts are made on the soil, disruption of its natural functions and of its historical use should be avoided as far as possible." Soil, within the meaning of this Act, is the upper layer of the earth's crust, as far as this layer fulfils the soil functions mentioned below, including its liquid (soil solution) and gaseous components (soil air), but excluding groundwater and the beds of water bodies. Its functions are:

Natural functions

- providing a basis for life and habitat for people, animals, plants and soil organisms
- forming part of natural systems, especially by virtue of its water and nutrient cycles
- providing a medium for decomposition, balance and restoration as a result of its filtering, buffering and substance converting properties, and especially groundwater protection

Functions useful to man

- a source of raw materials
- land for settlement and recreation
- land for agricultural use and forestry
- land for other economic and public uses, for transport, and for supply, provision and disposal

For the understanding of the Act there are some further definitions to be noted:

Harmful soil changes (adverse soil alterations) means harmful impacts on soil function that can result in hazards, significant disadvantage or nuisance to individuals or to the general public. This definition includes not only sites contaminated by active land use or industry, but also harmful landslides, erosion, compaction and others. A special group of harmful changes is referred to as *Altlasten,* meaning contamination on closed installations, or where current use is different from the historical one. In the definition of the Act are subsumed:

- closed waste management installations, and other properties in/on which waste has been treated, stored or land-filled (former waste disposal sites, Altablagerungen)
- properties that house closed installations, and others having handled substances which could cause harmful changes to the soil or present a hazard for individuals or the general public (former industrial sites, Altstandorte). This excludes installations which can be closed only under a license pursuant to the Atomic Energy Act.

Thus *remediation* refers to measures which:

- eliminate or reduce pollutants (i.e., decontamination)
- prevent or reduce the long-term spread of pollutants, without eliminating the pollutants themselves (i.e., securing containment)
- eliminate or reduce harmful changes to the soil's physical, chemical or biological characteristics

Protection and restriction measures are those which prevent or reduce hazards, significant disadvantage or nuisance to individuals or the general public, especially usage restrictions. Fences and warning signs belong within this definition.

The scope of application must be read in the context of Article 3 of the Act, as many limitations are defined where older Acts have regulated uses of land and soil such as fertilizing, construction modification, maintenance and operation of transport routes, searches for, recovery, transport, storage, treatment and destruction of military combat devices (including unexploded ordnance) and harmful changes to the soil caused by air pollution, noise, vibration, and similar phenomena (regulated in Article 3 of BimSchG, the Federal Act covering the prevention of harmful effects on the environment).

Table 1. Authorization to the Federal Government by the Federal soil Protection Act.

BBodSchG: Authorization to issue Ordinances to the Federal Government	
Article	
5	Unsealing of sealed ground
6	Adding of materials on to or into the soil
8 (1)	Trigger values, action values and requirements for warding off harmful changes and remediations
8 (2)	Precautionary reqirements and values
8 (3)	Procedures of identification of environmentally hazardous substances and reqirements to sampling, further treatment and quality control
13	Invrequirements for remediation investigations and remediating plans
23	Execution of the Act and relevant ordinances on military sites

The Act authorizes the Federal Government, after hearing from the parties concerned (Article 20) and with the agreement of the Chamber of the States (Bundesrat), to issue Ordinances (Table 1) to:

- oblige property owners to unseal sealed ground violating the provisions of construction law, and on land that is to remain unused in the long term whose sealing would violate decisions made under planning law (Article 5).
- meet requirements for the addition of materials onto or into the soil with respect to pollution content and other properties (Article 6).
- meet the provisions for the fulfilment of the above obligations arising from Article 4, for the investigation and assessment of suspect sites, harmful soil changes, sites suspected of being contaminated and contaminated sites. Such Ordinances may include, in particular, the definition of trigger values, which, if exceeded, induce further investigation and assessment whether there is a contaminated site or not; the definition of action values, the exceeding of which will normally signal the presence of a contaminated site and mean that measures are required; and to meet provisions for the prevention of harmful soil changes, including requirements related to the handling of excavated and treated soil material and for the remediation of the soil and contaminated sites, especially regarding the rehabilitation objective, the extent of decontamination measures and safeguarding measures, and the protection and restriction measures (Article 8 (1)).
- meet provisions for the fulfilment of precautionary obligations to soil conservation covered by Article 7 as well as for the requirements of relevant investigation and assessment of land where harmful soil changes are suspected to be occurring; to issue regulations with respect to precautionary values, the exceeding of which will confirm the risk of harmful soil change, and to issue regulations concerning additional pollution load, and requirements for the prevention or reduction of input of substances (Article 8 (2)).
- together with the values above, define possible values, procedures for determination of levels of environmentally harmful substances in soils, in biological materials and in other materials. Such procedures will include requirements for representative sampling, further treatment of samples, quality control, including the determination of values for different types of pollution. This task is not the responsibility of the Federal Government, the States and its proper authorities, or the consulting experts. However, the publication of procedures by the

Federal Government sets the standards for groups occupied with the Federal Soil Protection and the Contaminated Sites Ordinance.

- Regarding the requirements for remediation investigations and for the content of remediation plans (Article 13 (1)) to ensure that the Act and the Ordinances based on it are executed within the responsibility of the Federal Ministry of Defence and for the armed forces stationed in the Federal Republic of Germany as a result of agreements under international law by the Federal Ministry of Defence or of bodies chosen by this ministry.

The Act assigns to the States the following responsibilities:

- Proper authorities of the States may order unsealing measures in individual cases where no federal Ordinance has been issued and the provisions of Article 5 are respected (Article 5).
- The relevant proper authority shall take appropriate measures to determine the facts, if the authority is aware of indications of a harmful soil change or contaminated site, and where hazardous substance values exceed the relevant trigger values, the authority must take the necessary measures to determine whether or not there is a harmful soil change or a contaminated site, and to inform the owner or occupant of the property of the result of investigation and assessment, upon application, in writing (Article 9 (1)).
- If there are specific indications to suspect the presence of a harmful soil change or contaminated site, the proper authority may order the obliged persons to carry out necessary investigations and a risk assessment. The authority may require the carrying out of investigations and risk assessment by experts related to the requirements of Article 18 of the Act (Article 9 (2)).
- Proper authorities may take further measures pursuant to Articles 4 and 7, issue pursuant Ordinances on the basis of Articles 6 and 8, and where securing containment measures are necessary, demand security in advance and restrict agricultural and forest land use under specific conditions (Article 10).
- The States may issue provisions regarding the identification of contaminated sites and of sites suspected of being contaminated (Article 11).
- The proper authorities will require a remediation plan in the case of severely contaminated sites, which must be carried out by an expert relevant to Article 18, upon demand of the authority. The authority may alter the remediation (Article 13), or the authority may carry out the

remediation plan by itself, or alter or have it altered by experts relevant to Article 18 (Article 14).

- The proper authority surveys and supervises contaminated sites and suspected sites and may oblige the property owner etc. to undertake monitoring (Article 15).
- The proper authority may issue supplementary orders in fulfilment of obligations of part 3 of the "supplementary provisions for contaminated sites".
- The States may define further details for the requirements pertaining to experts and investigation agencies in order to achieve the necessary level of reliability and quality. They may define further details regarding the nature and extent of their tasks, the submission of the results of their activities and the publication of the names and addresses of experts who fulfil the relevant requirements.
- The States may issue supplementary procedural requirements for the implementation of the "principles and obligations" and the "supplementary provisions for contaminated sites" (Article 21).

The Act introduces many obligations to various persons, mainly to the landowner, as listed in Table 2.

3.2.2. *Federal Soil Protection and Contaminated Site Ordinance (BBodSchV)*

The Federal Government decreed this Ordinance in July 1999, after hearing from the parties concerned and with the agreement of the Bundesrat. The BBodSchV uses the authorizations of the Federal Soil Protection Act for regulations covering:

- the filling of materials onto or into the soil (pursuant to BBodSchG Article 6).
- the requirements and values for the identification, investigation and initial assessment of suspected sites and suspected contaminated sites (BBodSchG Article 8 (1), No. 1 and 2).
- the requirements and values for the prevention of hazards and the remediation of harmful soil changes and contaminated sites (BBodSchG Article 8 (1), No. 3).

Table 2. Obligated persons and relevant obligations by the German Federal Soil Protection Act.

Obligation	**Obligated Person or Party**	**Article of Act**
General obligation, that actions affecting soil shall not cause harmful soil changes	Everybody whose actions affect soil	4 (1)
Prevention measures against harmful soil changes originating from their property	Property owner Occupant of a property	4 (2)
Remediation of harmful soil changes and contaminated sites and of water pollutions caused by those in a manner that no hazards, considerable disadvantages or considerable nuisances for individuals or the general public occur, even in the long term If this is not possible or cannot be reasonably required, other protection or restriction measures shall be carried out	Person who caused a harmful soil change or contaminated site His universal successor Relevant property owner Occupant of the relevant property Person who , for reasons of commercial law or company law, are required to answer for a legal entity that owns a real property that is encumbered with harmful soil changes or site contamination; person who gives up ownership of such properties The former owner of a real property, if h has transferred his property after 1st of march 1999 and if he was aware of, or should have been aware of, the relevant harmful soil change or site contamination.	4 (3)
Precautionary measures in cases that actions can lead to harmful soil changes	Property owner Occupant over a site Party who carries out or gets carried out actions on a site	7

- the requirements and values for relevant investigation and assessment of pieces of land where there is concern that harmful soil changes are taking place, especially precautionary values and permissible additional pollution loads (BBodSchG Article 8 (3)).

This Ordinance is applied to:

- the investigation and evaluation of suspected sites, sites suspected of being contaminated, adverse soil alterations and contaminated sites, as well as to the requirements with respect to sampling, analysis and quality assurance pursuant to Article 8 (3) and Article 9 of the Federal Soil Protection Act.
- requirements with respect to risk prevention via decontamination and securing measures, as well as other protective and restrictive measures pursuant to Article 4 (2) to (5) and Article 8(1) second sentence No. 3 of the Federal Soil Protection Act.
- supplementary requirements with respect to investigations and plans for remediation for specific contaminated sites pursuant to Article 13(1) of the Federal Soil Protection Act.
- requirements concerning precautions against the formation of adverse soil alterations pursuant to Article 7 of the Federal Soil Protection Act, including the requirements with respect to the application and introduction of materials pursuant to Article 6 of the Federal Soil Protection Act.
- the determination of trigger and action values, as well as of precautionary values including the permissible additional pollution load pursuant to Article 8(1) second sentence Nos. 1 and 2 and paragraph (2) Nos. 1 and 2 of the Federal Soil Protection Act.

For better understanding, definitions are given for soil material, area of impact, exploratory investigation, detailed investigation, leachate forecast, pollutants, conditions of exposure, pathway, background content, erosion area, and root-penetrable soil layer. The following topics are presented in the Ordinance (the regulations concerning the investigation and evaluation of suspected sites and sites suspected of being contaminated are listed below):

- Investigation
- Indications of the existence of a contaminated site

- Scope of the investigation and required level of knowledge
- Investigations regarding the soil - human health pathway and types of land use
- Investigations regarding the soil - plant pathway, and the soil - groundwater pathway, exploratory investigations
- Detailed investigation
- Sampling
- Determination of sampling points and sampling depths for the soil - human health and soil - plant pathways, and the soil - groundwater pathway (Table 3)
- Soil gas sampling
- Preservation, transport and storage of samples
- Investigation methods for soils, soil material and other materials
- Selection and pretreatment of samples
- Extraction and elution techniques (Table 4)
- Methods for the analysis of soils, soil material and other materials
- Eluates and leachates (Tables 5, 6 and 8)
- Estimation of substance input into the groundwater
- Quality control: The standards, technical regulations and other methods, sources of supply are listed in No. 6 of annex 1
- Evaluation, using trigger values and action values (Tables 9 to 17)
- Remediation: investigation and planning
- Remediation measures, protective and restrictive measures, decontamination measures, securing measures
- Supplementary regulations for the prevention of risk of harmful soil changes by aquatic erosion
- Precaution requirements (Table 18)
- Requirements concerning filling materials onto or into the soil

Table. 3. related to Tab. 1 BBodSchV: Sampling depth depending on landuse.

Soil - plant pathways		
Pathway	Use	Sampling depth (cm)
soil - human health	playground, residential area	0-10 * 10-35 **
	park and recreational facility	0-10 *
	land used for industry and commerce	0-10 *
soil - plant	agriculture, vegetable garden	0-30 *** 30-60
	grassland	0-10**** 10-30

*) Contact area for oral and dermal pollutant intake, additional 0-2 cm in cases in which the inhalational intake is relevant

**) 0-35 cm: average thickness of applied soil layers; at the same time, maximum depth that can be reached by children

***) Working horizon

****) Primary rooting depth

Table 4. (Table 2 BBodSchV): Method for the preparation of eluates with water.

Method for the preparation of eluates with water Inorganic Substances		
Process	Specification	Method
soil saturation extract	for process see 1)	
elution with water	-sample mass in consideration of the dry matter according to DIN 38414-2: 11.85 or DIN ISO 11465:12.96 - for filtration see (2)_	DIN 38414-4: 10.84
organic substances		
column or lysimeter test	Note the speed with which the substance-specific equilibrium concentration establishes itself	

Table 5. (Table 3 BBodSchV): Analysis of physico-chemical properties.

Analysis of physico-chemical properties		
Investigation parameter	Specifications	Method
determination of the dry matter	soil samples fresh from the field or air-dried	DIN ISO 11465: 12.96
organic carbon and total carbon after dry combustion	air-dried soil samples	DIN ISO 10694: 08.96
pH-value (CaCl2)	Suspension of the soil sample fresh from the field or air-dried in CaCl; solution; c(CaCl2): 0.01 mol/1	DIN ISO 10390: 05.97
grain-size distribution	l) "finger test" in the field *	Pedological Mapping Guide, 4'" edition, 1994; DIN 19682-2: 04.97
	2) screening, dispersion, pipette analysis *	E DIN ISO 11277: 06.94 DIN 19683-2: 04.97
	3) screening, dispersion, areometer method	DIN 18123: 11.96 E DIN ISO 11277:06.94
Bulk density	drying of a soil sample taken in the proper volume at 105°C, back weighing	E DIN ISO 11272: 01.94 DIN 19683-12: 04.73

*Recommended methods

Table 6.a. (Table 5 BBodSchV , part 1) Analysis of organic pollutant concentrations.

Analysis of organic pollutant concentrations		
Investigation parameter	Specifications	Method
polycyclic aromatic hydrocarbons (PAH): 16 PAH (EPA) benzo(a)pyrene	l) Soxhiet extraction with toluene, Chromatographie clean-up; quantification by means of GC-MS* 2) extraction with tetrahydro-füran or acetonitrile; quantification by means of HPLC-UV/DAD/F* 3) extraction with acetone, adding petroleum ether, removal of acetone, Chromatographie clean-up of the petroleum ether extract, take-up in acetonitrile; quantification by means of HPLC-UV/DAD/F 4) extraction with a water/ acetone/petroleum ether mix in the presence of NaCl; quantification by means of GC-MS or HPLC-UV/DAD/F	Guidelines No. l of LUA-NRW, 1994* Guidelines No. l of LUA-NRW, 1994* E DIN ISO 13877: 06.95 VDLUFA Book of Methods, vol. VII; Contaminated Sites Manual vol. 7 LfU HE
Hexachlorobenzene	extraction using acetone/cyclo-hexane mix or acetone/petroleum ether; if necessary, Chromatographie clean-up after removal of the acetone; quantification by means of GC-ECD or GC-MS	E DIN ISO 10382: 02.98
Pentachlorophenol	Soxhiet extraction using heptane or acetone/heptane (50:50); derivatisation with acetic anhydride; quantification by means of GC-ECD or GC-MS	E DIN ISO 14154: 10.97

Table 6.b. (Table 5 BBodSchV, part 2): Analysis of organic pollutant concentrations.

Analysis of organic pollutant concentrations		
Investigation parameter	Specifications	Method
aldrin, DDT, HCH-mix	l) extraction using petroleum ether or acetone/petroleum ether mix, Chromatographie clean-up; quantification by means of GC-ECD or GC-MS* 2) extraction using water/acetone/ petroleum ether mix, quantification by means of GC-ECD or GC-MS	E DIN ISO 10382:02.98* a VDLUFA Book of Methods, vol. VII
polychlorinated biphenyls (PCB): 6 PCB-congeners (No. 28, 52, 101, 138,153,180 according to Ballschmiter)	l) extraction using heptane or acetone/petroleum ether, Chromatographie clean-up, quantification by means of GC-ECD (GC-MS possible) 2) Soxhiet extraction using heptane, hexane or pentane, Chromatographie clean-up involving AgN03/silica column; quantification by means of GC-ECD (GC-MS possible) 3) extraction with a water/ acetone/petroleum ether mix in the presence of NaCl; quantification by means of GC-ECD (GC-MS possible)	E DIN ISO 10382:02.98* DIN 38414-20: 01.96 VDLUFA Book of Methods, vol. VII
polychlorinated dibenzodioxins and dibenzofürans	Soxhiet extraction of freeze-dried samples with toluene, Chromatographie clean-up; quantification by means of GC-MS	according to the Sewage Sludge Ordinance and in consideration of DIN 38414-24: 04.98, VDI Guideline 3499, sheet *l:* 03.90

* recommended method

Table 7. (Table 6 BbodSchV): Determination of the concentration of inorganic pollutants in eluates and leachate.

Determination of the concentration of inorganic pollutants in eluates and leachate		
Investigation parameter	Specifications	Method
As, Cd, Cr, Co, Cu, Mo, Ni, Pb, Sb, Se, Sn, Tl, Zn	ICP-AES (ICP-MS possible)	on the basis of DIN EN ISO 11885:04.98*
arsenic (As), antimony (Sb)	hydrid AAS	DIN EN ISO 11969: 11.96
Lead (Pb)	AAS	DIN 38406-6: 07.98
cadmium (Cd)	AAS	DIN EN ISO 5961: 05.95
chromium (Cr), total	AAS	DIN EN 1233: 08.96
chromium (Cr VI)	spectrophotometry ion chromatography	DIN 38405-24: 05.87 DIN EN ISO 10304-3: 11.97
cobalt (Co)	AAS	DIN 38406-24: 03.93
copper (Cu)	AAS	DIN 38406-7: 09.91
nickel (Ni)	AAS	DIN 38406-11:09.91
mercury (Hg)	AAS vapor compression technique	DIN EN 1483: 08.97
selenium (Se)	AAS	DIN 38405-23: 10.94
Zinc (Zn)	AAS	DIN 38406-8: 10.80
cyanide (CN-), total	spectrophotometry	DIN 38405-13: 02.81 E DIN EN ISO 14403: 05.98
cyanide (CN-)	spectrophotometry	DIN 38405-13: 02.81
fluoride (F-)	fluoride-sensitive electrode ion chromatography	DIN 38405-4: 07.85 DIN EN ISO 10304-1: 04.95

*The determination limit must be adapted to the examination target by suitable measures or suitable technical equipment.

Table 8. (Table 7 BBodSchV): Determination of the concentration of organic pollutants inthe soil leachate.

Determination of the concentration of organic pollutants in the soil leachate		
Investigation parameter	Specifications	Method
benzene	GC-FID	DIN 38407-9: 05.91*
BTEX	GC-FID matrix load must be complied with	DIN 38407-9: 05.91
highly volatile halogenated hydrocarbons	GC-ECD	DIN EN ISO 10301: 08.97
aldrin	GC-ECD (GC-MS possible)	DIN 38407-2: 02.93
DDT	GC-ECD (GC-MS possible)	DIN 38407-2: 02.93
phenols	GC-ECD	ISO/DIS 8165-2: 01.97
chlorophenols	GC-ECD or GC-MS	ISO/DIS 8165-2: 01.97
chlorobenzenes	GC-ECD (GC-MS possible)	DIN 38407-2: 02.93
PCB, total	GC-ECD GC-ECD or GC-MS	DIN EN ISO 6468: 02.97 ,DIN 51527-1: 05.87 DIN 38407-2: 07.98
PAH, total	HPLC-F	DIN 38407-8: 10.95
naphthalene	GC-FID or GC-MS	DIN 38407-9: 05.91
petroleum hydrocarbons	extraction with petroleum ether, gas Chromatographie quantification	according to ISO/TR 11046: 06.94

* the determination limit must be adapted

Table 9. (BBodSchV – Annex 2, p. 51): Soil – Human Health Pathway – Trigger Values.

Soil – Human Health Pathway (Direct Contact)				
Trigger values [mg/kg TM]				
Substance	Playgrounds	Residential areas	Parks and recreational facilities	Land used for industrial and commercial purposes
Arsenic	25	50	125	140
Lead	200	400	1,000	2,000
Cadmium	10*	2*	50	60
Cyanides	50	50	50	100
Chromium	200	400	1,000	1,000
Nickel	70	140	350	900
Mercury	10	20	50	80
Aldrin	2	4	10	—
Benzo(a)pyrene	2	4	10	12
DDT	40	80	200	—
Hexachlorobenzene	4	8	20	200
Hexachlorocyclo-hexane (HCH-mix or (3-HCH)	5	10	25	400
Pentachlorophenol	50	100	250	250
Polychlorinated biphenyls (POP«)**	0.4	0.8	2	40

*) In back gardens and small gardens where children stay and food plants are grown, the trigger value 2.0 mg/kg TM must be applied in the case of cadmium.

**) Where PCB total contents are determined, the measured values must be divided by a factor of 5

Table 10. (Table BBodSchV annex 2 – 1.3): Soil – Human health pathway; Action Values.

Soil – Human health pathway Action values (ng I-TEq/kg TM)*)				
Substance	Play-grounds	Residential areas	Parks and recreational facilities	Land used for industrial and commercial purposes
Dioxins/furanes (PCDD/F)	100	1 000	1 000	10 000
*) Sum of the 2,3,7,8-TCDD-toxicity equivalents (according to NATO/CCMS).				

Table 11. (BBodSchV – Annex 2, 2.4, p. 53): Soil – Plant Pathway; Agriculture, with regard to growth impairments of cultivated plants.

Soil – Plant Pathway	
Agriculture with regard to growth impairments of cultivated plants	
Trigger values	
Substance	Trigger value (mg/kg dry matter, fine soil, in ammonium nitrate extract)
Arsenic	0.4
Nickel	1.5
Copper	1.0
Zinc	2.0

Table 12. (BBodSchV annex 2, 2.2, p.52): Soil – Plant Pathway.

Soil – Plant Pathway			
Action and trigger values (mg/kg dry matter, fine soil)			
Agriculture, vegetable garden			
Substance	Method*	Trigger value	Action value
Arsenic	KW	200**	—
Cadmium	AN	—	0.04/0. 1***
Lead	AN	0.1	—
Mercury	KW	5	
Thallium	AN	0.1	—
Benzo(a)pyrene	—	1	—

*) Extraction process for arsenic and heavy metals: AN - ammonium nitrate, KW = aqua regia (Königswasser)

**) In the case of soils with temporarily decreasing conditions, a trigger value of50 mg/kg dry matter must be applied.

***) In areas that are used for growing bread wheat or strongly cadmium-accumulating vegetables, an action value of 0.04 mg/kg dry matter must be applied; otherwise, the action value is 0.1 mg/kg dry matter 2.3 Action values (pursuant to Artiele 8 (l) second sentence No. 2 of the Federal Soil Protection Act) in relation to plant quality for the pollutant transition soil - plant on grassland areas (in mg/kg dry matter, fine soil, arsenic and heavy metals in aqua regia extract, analysis according to Annex l)

Table 13. (BBodSchV annex 2, 2.3, p.53): Soil – Plant Pathway; Action Values; Grassland.

Soil – Plant Pathway	
Action values (mg/kg dry matter, fine soil)	
Grassland	
Substance	Action value
Arsenic	50
Lead	1 200
Cadmium	20
Copper	1 300 1)
Nickel	1 900
Mercury	2
Thallium	15
Polychlorinated biphenyls (PCB6)	0.2

1) Where sheep are kept on grassland, the applicable action value is 200 mg/kg dry matter

Table 14. (Table BBodSchV annex 2 – 1.3): Soil – Human Health Pathway; Trigger Values.

Soil – Human Health Pathway				
Action Values (ng I-TEq/kg TM)*)				
Substance	Play-grounds	Residential areas	Parks and recreational facilities	Land used for industrial and commercial purposes
Dioxins/furanes (PCDD/F)	100	1 000	1 000	10 000
*) Sum of the 2,3,7,8-TCDD-toxicity equivalents (according to NATO/CCMS).				

Table 15. (BBodSchV – Annex 2, 2.4, p. 53): Soil – Plant Pathway; Agriculture; with regard to growth impairments of cultivated plants Trigger Values.

Soil – Plant Pathway	
Agriculture with regard to growth impairments of cultivated plants	
Trigger values	
Substance	Trigger value (mg/kg dry matter, fine soil, in ammonium nitrate extract)
Arsenic	0.4
Nickel	1.5
Copper	1.0
Zinc	2.0

Table 16. (BBodSchV annex 2, 2.2, p.52): Soil – Plant Pathway; Action and Trigger Values; Agriculture, vegetable garden.

Soil – Plant Pathway			
Action and Trigger Values (mg/kg dry matter, fine soil)			
Agriculture, vegetable garden			
Substance	Method*	Trigger value	Action value
Arsenic	KW	200**	—
Cadmium	AN	—	0.04/0. 1***
Lead	AN	0.1	—
Mercury	KW	5	
Thallium	AN	0.1	—
Benzo(a)pyrene	—	1	—

*) Extraction process for arsenic and heavy metals: AN - ammonium nitrate, KW = aqua regia (Königswasser)

**) In the case of soils with temporarily decreasing conditions, a trigger value of50 mg/kg dry matter must be applied.

***) In areas that are used for growing bread wheat or strongly cadmium-accumulating vegetables, an action value of 0.04 mg/kg dry matter must be applied; otherwise, the action value is 0.1 mg/kg dry matter 2.3 Action values (pursuant to Artiele 8 (l) second sentence No. 2 of the Federal Soil Protection Act) in relation to plant quality for the pollutant transition soil - plant on grassland areas (in mg/kg dry matter, fine soil, arsenic and heavy metals in aqua regia extract, analysis according to Annex l)

Table 17. (BBodSchV annex 2, 2.3, p.53): Soil – Plant Pathway; Action Values; grassland.

Soil – Plant Pathway	
Action Values (mg/kg dry matter, fine soil)	
Grassland	
Substance	Action value
Arsenic	50
Lead	1 200
Cadmium	20
Copper	1 300 1)
Nickel	1 900
Mercury	2
Thallium	15
Polychlorinated biphenyls (PCB6)	0.2

Where sheep are kept on grassland, the applicable action value is 200 mg/kg dry matter

Table18. Precautionary Values.

Precautionary values for soils pursuant to Article 8 (2) No. l of the Federal Soil Protection Act (analysis according to Annex l BBodSchV) Precautionary values for metals (mg/kg dry matter, fine soil, aqua regia decomposition)							
Soils/Soil type	Cadmium	Lead	Chromium	Copper	Mercury	Nickel	Zinc
clay	1.5	100	100	60	1	70	200
loam/silt	1	70	60	40	0.5	50	150
sand	0.4	40	30	20	0.1	15	60
Soils with naturally increased and settlement-related increased background concentrations over large areas	safe, provided that the release of pollutants or additional inputs pursuant to Article 9 (2) and (3) of this Ordinance do not give reason to expect any adverse impacts on the soil functions						

Precautionary values for organic substances (mg/kg dry matter, fine soil)			
Soils	Polychlorinated biphenyls (PCB,,)	Benzo(a)pyrene	Polycyclic aromatic hydrocarbons (PAH)«)
Humus content > 8 %	0.1	1 0.3	10 3
Humus content < 8 %	0.05	1 0.3	10 3

4. Conclusions

The Soil Protection Act and Ordinance unify the regulations covering the detection, sampling and analysis of contamination, and establish common measures for contamination assessment and remediation within the Federal Republic of Germany. These regulations may be suitable for regions which currently lack such detailed legal provisions to a common standard within the Federal Republic of Germany. However, the Act and the Ordinance are embedded in a network of older laws which regulate important sectors of soil protection, such as the planning of land use. In this respect, the influence of the Act is limited. It is possible that the EU will be able to improve situations where the applicability of the German Soil protection law is limited, and help to develop anticipatory and complete soil protection law. Finally it should be noted that there are still possibilities to simplify the regulations governing the application of biological remediation, and to find better means of funding the investigation and assessment of precautionary and remedial measures for contaminated soil.

References

Federal Government (1998): Federal Soil Protection Act ; Federal Law Gazette I p. 502 – 510, German

Federal Government (1999): Federal Soil Protection and Contaminated Site Ordinance (BBodSchV); Federal Law Gazette I, p. 1554 – 1582, German

Becker, B. (2005): Federal Soil Protection Act, Comment; R.S. Schulz GmbH, Starnberg, German

Holzwarth, E. (2000): Federal Soil Protection Act, Comment; 2nd edition, Berlin, 2000, German

Sanden, J. (1998): Federal Soil Protection Act, Short Comment; Heidelberg 1998, German

Versteyl et. al. (2002): Federal Soil Protection Act, Comment; München 2002, German

THE MAIN DIRECTIONS OF COOPERATION BETWEEN THE MINISTRY OF NATURAL RESOURCES OF THE RUSSIAN FEDERATION AND CCMS NATO WITHIN THE FRAMEWORK OF NRC CCMS

A.L. KOZELTSEV *
Adviser to the Minister of Natural Resources of the Russian Federation, national coordinator to CCMS NATO

*To whom correspondence should be addressed.

The NRC Committee on the Challenges of Modern Society (CCMS) was established by the NATO-Russia Council (NRC) according to the provisions of the "Declaration by Heads of State and Government of NATO Member States and the Russian Federation", adopted in Rome on 28 May, 2002, building on the Founding Act on Mutual Relations, Cooperation and Security between NATO and the Russian Federation. The aim of the NRC CCMS is to promote, encourage and coordinate joint co- operative projects involving experts from NATO countries and Russia on new threats and challenges to security, including environmental protection problems arising from civilian and military activities, as well as topics of primary importance to both parties as formulated by the NATO-Russia Council. The Ministry of Natural Resources is the main partner of CCMS NATO with respect to environmental protection problems within the framework of NRC CCMS. In 2003 a list of twelve topics was presented by the Ministry of Natural Resources of the Russian Federation to CCMS NATO and then adopted as a Mutual Plan of Actions.

These topics are:

14. Training and advanced training courses for military and civil environmentalists in the field of environmental protection and environmental safety
15. Environmentally-friendly industrial technologies

N. Marmiroli et al. (eds.), Advanced Science and Technology for Biological Decontamination of Sites Affected by Chemical and Radiological Nuclear Agents, 257–258.

16. Environmental rehabilitation of sites (including the sites of former military units) polluted as a result of military activity
17. Environmental safety measures when treating hazardous wastes
18. The problems of environmental safety of rocket and space activity
19. Improvement of management in the field of the environmental protection and safety
20. Measures for improving the quality of water, including drinking water, in the water supply system
21. Measures for prevention of ecological terrorism and elimination of its consequences
22. Ecological aspects of peacekeeping operations
23. Environmental safety measures in relation to methods and technologies, and organising work on the prevention and elimination of the consequences pollution of nature ecosystems with oil products
24. Ecological aspects of the destruction and utilization of armaments and military equipment
25. General methodology and criteria for ecological risk and standards

The short-term project, jointly led by Italy and the Russian Federation, on the development of a prototype system for sharing information related to acts of terrorism to the environment, agriculture and water systems is aimed to protect these systems from possible attacks and to minimize the effects if attacks do occur. The project focuses primarily on developing a database system as a tool for gathering, organising and evaluating information. This includes categories, such as threat agents, threat scenario of concern, prevention and security measures and preparedness plan. This project, so called "ECOTER" was adopted at NRC CCMS plenary in April, 2005.

SUBJECT INDEX

9 781402 055188

MIX
Papier aus verantwortungsvollen Quellen
Paper from responsible sources
FSC® C105338

If you have any concerns about our products,
you can contact us on
ProductSafety@springernature.com

In case Publisher is established outside the EU,
the EU authorized representative is:
Springer Nature Customer Service Center GmbH
Europaplatz 3, 69115 Heidelberg, Germany

Printed by Libri Plureos GmbH
in Hamburg, Germany